Palgrave Advances in the Economics of Innovation and Technology

Series Editor
Albert N. Link, Department of Economics, University of North
Carolina at Greensboro, Greensboro, NC, USA

The focus of this series is on scholarly inquiry into the economic foundations of technologies and the market and social consequences of subsequent innovations. While much has been written about technology and innovation policy, as well as about the macroeconomic impacts of technology on economic growth and development, there remains a gap in our understanding of the processes through which R&D funding leads to successful (and unsuccessful) technologies, how technologies enter the market place, and factors associated with the market success (or lack of success) of new technologies.

This series considers original research into these issues. The scope of such research includes in-depth case studies; cross-sectional and longitudinal empirical investigations using project, firm, industry, public agency, and national data; comparative studies across related technologies; diffusion studies of successful and unsuccessful innovations; and evaluation studies of the economic returns associated with public investments in the development of new technologies.

T. S. Krishnan

Untapped Knowledge in India's E-Waste Industry

A Roadmap to Strengthen the Informal Economy

T. S. Krishnan
Birmingham, UK

ISSN 2662-3862   ISSN 2662-3870 (electronic)
Palgrave Advances in the Economics of Innovation and Technology
ISBN 978-3-031-50295-8   ISBN 978-3-031-50296-5 (eBook)
https://doi.org/10.1007/978-3-031-50296-5

Cover illustration: © John Rawsterne/patternhead.com

This Palgrave Macmillan imprint is published by the registered company Springer Nature Switzerland AG
The registered company address is: Gewerbestrasse 11, 6330 Cham, Switzerland

Paper in this product is recyclable.

PREFACE

This book is a synthesis of the core elements of my 780-page long PhD thesis: "Understanding E-Waste Reverse Supply Chain and Formalization of Informal E-Waste Processors: A Qualitative Case Study," which I completed in Production and Operations Management area at the Indian Institute of Management (IIM) Bangalore.

My doctoral research journey spanned more than eight years. It was partly funded by the Ph.D. fellowship from IIM Bangalore and partly funded by my wife. The research included 2 years of intense ethnographic fieldwork during which I interviewed more than 90 industry experts. I call my methodology *ethnographic* because it centers on developing a deep understanding of people and context by *immersing* in the industry.

When I began this research, my goal was to incentivize informal e-waste recyclers to become formal. I was thinking of how to make them comply with the prevailing e-waste law. But my ethnographic fieldwork led me in the opposite direction, and I began to question the foundational assumptions of formalization.

I believe my research and this book represents operations management scholarship as defined by Kostas Dervitsiotis in his classic textbook Operations Management, published in 1981:

It is the study of a field where knowledge from the social sciences (economics, psychology, sociology), engineering, and mathematics comes together to create and operate productive systems for the satisfaction of human needs in products and services.

While researching India's e-waste recycling industry, I realized that I knew very little about India's informal economy and how local industries work. I also realized how politics, history, culture, and society influence how the industry functions. To understand and synthesize what I saw in the field, I expanded my research beyond my home discipline of operations management and wove in learnings from economic history, sociology, development studies, archaeology, anthropology, environmental politics, critical geography, and other fields. This multidisciplinary approach led me to a more holistic understanding of India's e-waste recycling industry.

It took me five years to reframe and rewrite my specialized PhD thesis into the book you are currently reading. This book is written in an accessible manner to reach a wider audience of scholars and practitioners of waste management and sustainable development. The book is best understood if it is read sequentially from Chapters 1 to 5 and reflecting on these chapters as a whole.

I hope this book sparks the reader's interest in three ways:

- Encourage future operations management scholars to do ethnographic fieldwork and multidisciplinary research on *messy* topics and ask *big questions* to advance scholarship.
- Influence public discussions of India's e-waste policy by challenging the dominant bias toward Extended Producer Responsibility.
- Provide pathways to develop contextual solutions to manage e-waste in the Global South and foster a thriving circular economy across industries.

Birmingham, UK   T. S. Krishnan

ACKNOWLEDGMENTS

This book is dedicated to the 90+ experts from the e-waste recycling industry who shared their time and knowledge with me. They did this purely for goodwill, without expecting anything in return. I owe them a debt of gratitude. To protect their confidentiality, I am not calling out the names of these experts.

I am grateful to Pooja, my wife, for being the breadwinner and funding my doctoral research for more than three years. She also supported me when I took a six-month break from work to write this book full-time.

I thank Meera Seth and Bronwyn Geyer, commissioning editors at Palgrave Macmillan, for handling the entire process swiftly. Meera Seth was kind enough to grant me multiple extensions to meet deadlines that helped balance my family and professional commitments.

I thank the two anonymous reviewers for their constructive feedback that helped to improve this manuscript.

I thank the late N. S. Ramaswamy, founder director of IIM Bangalore, for encouraging me to pursue research on India's waste management industry.

I am grateful to Barbara Harriss-White, professor at the University of Oxford, for taking the time to listen to my research and encouraging me to pursue public scholarship. It was exciting to dream with Barbara about a multidisciplinary waste institution affiliated with a prestigious university.

Many colleagues and faculty have helped me during my Ph.D. journey and this book publication. I would like to call out specific individuals who went beyond the call of their duty by providing encouragement, moral support, compassion, and a job during the final and sometimes challenging stages of my Ph.D. research: Rajiv Kozhikode, Krishna Sundar, Nayana Tara, Rejie George, Ganesh Prabhu, Shooj Bhaskaran, Papi Reddy, Srikanth Krishnaprasad, Parvathy B, Ganesh Kumar, Damini Gupta, Jai Ganesh, Amol Agrawal, Althaf Shajahan, Srinidhi Raghavendra, and Shubham Singh.

Special thanks to my friendly colleagues at Mphasis NEXT Labs for their kindness and guidance as I rejoined into the corporate world and expanded my knowledge of bleeding-edge technologies.

Erica Machulak helped me to think through and structure my complex 780-page thesis, transitions between the components of my key findings, and how best to leverage my field evidence. This high-level framing to connect the dots helped me clarify what I wanted to get across in this book. I wish Erica was a member of my doctoral advisory committee. Simangele Mabena supported my writing process, and this helped to move my book forward by writing in small increments and celebrate wins, large and small.

PRAISE FOR *UNTAPPED KNOWLEDGE IN INDIA'S E-WASTE INDUSTRY*

"Krishnan's research makes a compelling case for designing e-waste policies in collaboration with informal processors by recognizing and valuing
their system. A similar case can, and should, be made for the informal
economy in general, not just in the e-waste sector or in India. This is an
important and timely book!"

*—Martha Alter Chen, International Coordinator, WIEGO; Lecturer
in Urban Planning and Design at the Graduate School of Design,
Harvard University*

"For decades policy that ignores its own necessary preconditions has
generated unintended side effects, as demonstrated by the thunder-
claps of demonetisation, GST, labour and farm law reforms. So recent
high-stakes policies for the formalisation of India's huge informal
economy have become contentious. Krishnan's study of e-waste, its
informal recycling and circular reprocessing - and of new policy to
formalise it - provides granular evidence of unintended policy effects and
of the paradoxical waste of the very human and material resources that the
policy seeks to develop. A must-read for wastistas and for policy scholars
and practitioners."

*—Barbara Harriss-White, Professor Emeritus of Development Studies,
Emeritus Fellow of Wolfson College, and Senior Research Fellow, School of
Interdisciplinary Area Studies, University of Oxford*

"Most of India's e-waste is recycled by the so-called informal processors, a highly functional network of specialists, who handle e-waste as small family businesses. India's current e-waste law based on Extended Producer Responsibility (EPR), favours formal processors, and creates barriers for informal processing. Its implementation has led to chaotic and economically inefficient scenarios. Krishnan's research argues for restructuring the EPR-based law to account for the ground reality. More generally, it raises questions about both the informal (and occasionally unsafe), and the formal (and often bureaucratic), approaches to waste management in the Global South, where a large fraction of the world's waste may have to be processed in the coming decades."

—Nitin Joglekar, *Associate Professor of Operations and Technology Management, Boston University and Departmental Editor*, Production and Operations Management

"This is a groundbreaking book that challenges conventional thinking and sheds light on the untapped potential within India's electronic waste industry. Krishnan's meticulous research and insightful analysis reveal how the informal sector, which has been thriving for centuries, embodies the principles of a circular economy. This work not only exposes the flaws in formalization policies but also makes a compelling case for leveraging the strengths of informal processors to drive economic and environmental gains. It is a must-read for those interested in sustainable development and e-waste management in India and other emerging economies. Additionally, the book would be a valuable resource for doctoral-level methods courses, serving as an excellent example of immersive field research."

—Rajiv Kozhikode, *Associate Professor for International Business and Management & Organization Studies, Beedie School of Business, Simon Fraser University*

"Informal sector e-waste processing in India is shrouded by secrecy due to the many illegalities involved. It is in this difficult and complex context that Krishnan has conducted a rare and intensive field study to learn how the informal sector has created knowledge to become effective and efficient while the formal sector is plagued by losses. Highly recommended for policy makers, managers, and researchers."

—Ganesh N. Prabhu, *Professor of Strategy, Indian Institute of Management Bangalore*

"Dr. Krishnan's research extends the boundaries of knowledge in sustainable operations management by including rich operational details from India's e-waste recycling industry and discussing how an appropriate public policy needs to be crafted. The research methodology exemplifies the best practices for conducting qualitative field-based operations management research that integrates theories from multiple disciplines."

—Kalyan Singhal, *McCurdy Professor of Business, University of Baltimore; Editor-in-Chief,* Production and Operations Management and *Founder and Co-Editor-in-Chief,* Management and Business Review

"This book offers unique insights into e-waste management and guides policy makers in recognizing the highly functional systems of informal processors. It is a must-read book for those working on e-waste, which is the need of the hour."

—S. Nayana Tara, *Former Professor, Centre for Public Policy, Indian Institute of Management Bangalore*

"This book makes the important point that attention to how informal economies function is crucial for successful recycling policies. It is a valuable contribution to sustainability policy studies."

—Carl Zimring, *Professor of Sustainability Studies, Pratt Institute; author of* Cash for Your Trash

About This Book

This research is based on the author's rare interviews with hard-to-reach stakeholders, including informal processors with whom the author has developed trust relationships over the course of years. This work draws upon the author's 90+ interviews with 49 stakeholders in India's e-waste processing industry over a period of two years. These stakeholders include large companies who generate e-waste, informal and formal processors, manufacturers, retailers, government, non-governmental organizations (NGOs), establishments in the second-hand market, waste incinerators, commodities (metals, plastics) traders and recyclers, and precious metal refiners. This unique primary research is corroborated and augmented by secondary sources including media coverage, white papers, and government documents. As such, the key findings explained in the book are based on qualitative (textual) evidence.

The actual names of people and organizations are disguised to protect the confidentiality of participants.

Legend of pseudonyms frequently used in the book

Pseudonyms	Description
Alpha, Beta, Gamma, Delta, Epsilon, Zeta, Mash	Informal-turned-formal e-waste processing companies
Aleph, Beet, Cool	Informal e-waste processing companies
RTH, Amy, EP, CIT	Formal e-waste processing companies
BB	Formal e-waste collection company
RWM	Machinery manufacturer for metal recovery from e-waste
Stimulus	EU-based international development agency
Consultant X, Consultant Y	Consultants with deep knowledge of India's waste management industry; both were employed by Stimulus to help formalize informal processors.
RML, LDR	Commodity recyclers
Compost	Waste management NGO that works closely with the informal processors
MMT	Scrap metal importer

A note on the interview quotes: The quotes cited in the book are original, and I have made editorial decisions to maintain the authenticity of respondents' voices. Except where indicated, I have not edited or paraphrased quotations provided by respondents in English. I have translated quotations provided in Hindi—the language spoken by many respondents. Square brackets next to certain words within the cited quotations denote the actual meaning of those words.

CONTENTS

Abbreviations

CPU	Central Processing Unit of a computer
CRT	Cathode-Ray Tube
EPR	Extended Producer Responsibility
EU	European Union
IT	Information Technology
LME	The London Metal Exchange
PCB	Printed circuit board
RoHS	The Restriction of Hazardous Substances
SDG	Sustainable Development Goals
TSDF	Treatment, Storage, and Disposal Facility
UNEP	United Nations Environmental Program
WEEE Directive	The Waste Electrical and Electronic Equipment Directive

Repositioning the "Informal Economy": The Hidden Potential for a New Circular Economy in India's Electronic Waste Industry

Abstract 80% of India's e-waste is recycled by its so-called *informal processors*, the highly-functional network of specialists who have been handling specialized waste for centuries. In the year 2010, responding to global policy trends and suggestions given by international NGOs, India's government introduced e-waste legislation to regulate the preexisting e-waste processing industry that was naturally structured to create value from e-waste. The push to formalize the informal processors through the e-waste law undermined a functional system. The flaws in this system become clear through an overview of the formalization process, the obstacles that formalization introduces, and the unanticipated challenges of formalization policies. A closer look at their practices and processes demonstrates that informal e-waste processors have long embodied traits that we associate with a thriving circular economy. The narrative that emerges offers a counterpoint to mainstream arguments that dismiss the value of informal processors.

Keywords E-waste recycling · Informal sector · Extended producer responsibility · E-waste legislation · Circular economy · Informal recycling · Formalization

INTRODUCTION

I first met Azhar in the summer of 2011 at his small, 200-square-foot electronics scrap shop in Bengaluru's scrap recycling neighborhood. He agreed to meet me based on the strong referral from the founder of a nongovernment organization. After warmly welcoming me to his shop, he shared his entrepreneurial story in the e-waste business. Azhar had been trained to reuse and repurpose e-waste as an apprentice with a family relative who was an expert in this field. For more than a decade, he slogged and learned the skills to reuse and repurpose (i.e., process) e-waste in new ways. Later, he began his own scrap shop, where we met, to process e-waste. Businesses like Azhar's are considered "informal" because they are not specifically approved by the government for e-waste processing.

Everything changed for Azhar when, in 2010, the government of India turned the e-waste industry on its head. Responding to global policy trends, India introduced a new law based on Extended Producer Responsibility (EPR) requiring e-waste processing businesses like Azhar's to set up facilities with government approval to process e-waste.[1] With the push from the newly developed e-waste law to *formalize* such informal e-waste processing businesses, Azhar started the *formal* company Alpha Recycling with an investment of ₹20,00,000 ($25,000 in today's market[2]). His company, Alpha Recycling, a 1000-square-foot facility, had been established recently in an area earmarked for industrial manufacturing. Alpha is one of the first informal-turned-formal e-waste processing companies in India. Azhar was proud of this achievement, and he was brimming with ideas to scale his business with the newly formed company. Azhar leveraged his social network and forged a deal to buy e-waste from a global technology services company. This single deal helped sustain and grow his business.

I met Azhar again five years later, in the summer of 2016. Alpha Recycling was temporarily closed, so we met in a crowded local tea shop in the same scrap recycling neighborhood. His enthusiasm and ideas to scale business were replaced with regret for having formalized. Azhar's ideas for scaling his business had not worked out and, gradually, the cost of running his business had increased fivefold. He no longer had the capital

---

[1] E-waste (Management and Handling) Rules were drafted in 2010 and made binding from 2011.

[2] Based on the average exchange rate in 2022.

nor the access to buy and process e-waste. Before formalizing, Azhar used to buy e-waste from global technology companies by paying higher prices and outperforming the strongest competitors in the formal sector. Now, he found that global companies preferred to sell to those who spoke good English, wore power suits, and prepared beautiful PowerPoint presentations. Azhar shared his frustration with me [slightly paraphrased for readability]:

> I have met lot of people from India and abroad for the last 10 years. The only discussion has been how to bring the informal sector to formal. I'm tired and fed up after hearing this same thing again and again. You are not thinking how to develop the existing informal sector as an industry in itself and help that industry grow. I have never seen such a project in my life.

A few months after he transitioned to set up the *formal* company, Azhar began closing his operations most days. Without the connections to procure e-waste from global companies, he was finding it hard to make ends meet. The company continues behind the scenes, and e-waste is reused and repurposed informally. Within the current system, this is the only way that Azhar can sustain his business and raise his family.

Azhar's experience is not an isolated one. There are many such *informal* e-waste processors[3] like Azhar who went on to set up *formal* companies like Alpha to comply with the e-waste law. Many community-based e-waste processors were drawn to formalization because of the promise of revenue growth and the potential to scale their businesses. They were presented with the idea of formalization as a pathway to global recognition, incentives from government, and a larger network of e-waste supply that could scale their businesses and achieve higher revenues. The tradeoff, it soon became clear, was small business owners' significant up-front investment in infrastructure that, for many, has proven to be ineffective.

In practice, informal processors who pursue formalization confront systemic hurdles that prevent them from developing scalable business models. They are unable to access higher quantities of e-waste to scale

---

[3] The processors, whether they are informal or formal, are commonly called as *recyclers*. The more appropriate term *processors* is used throughout this book, as it is more holistic and captures a wide range of operations that include reuse, repair, refurbish, cannibalize, disassembly, and metal recovery (Thierry et al., 1995).

their businesses. The complex administrative process of dealing with the government drains their energy. It is not necessarily the case that informal processors who formalize achieve better economic outcomes. From the perspective of informal processors, there is a false promise at play related to the benefits that formalization will reap. This false promise is based on an incomplete understanding of the economic, cultural, and political factors that make up India's e-waste processing industry.

To understand these systemic challenges, I conducted more than 90 interviews across the 49 stakeholders within India's e-waste processing industry. These stakeholders include large companies who generate e-waste, informal and formal processors, manufacturers, government, non-governmental organizations (NGOs), establishments in the second-hand market, waste incinerators, commodities traders and recyclers, and precious metal refiners. These conversations shed light on how India's current e-waste law has undercut its local talent and missed opportunities to leverage homegrown knowledge, human capital, and existing systems. These are rare interviews with hard-to-reach stakeholders, including informal processors with whom I have developed trust relationships over the past decade. This unique primary research is corroborated and augmented by secondary sources including recent scholarship, media coverage, white papers, and government documents. The narrative that emerges offers a counterpoint to mainstream arguments that dismiss the value of informal processors' expertise.

In this chapter, I explain the true meaning of *informal* and *formal* and highlight that the definition of *informal* is fluid in the context of India's e-waste industry. The flaws in this system become clear through an overview of the formalization process, the obstacles that formalization introduces, and the unanticipated challenges of formalization policies. This book argues that policies that leverage the strengths of informal processors are better placed to achieve sustainable outcomes. As the following chapters will show, minor policy adjustments enable solutions that reinforce and leverage the strengths of these informal processors and achieve global agendas such as circular economy and United Nations Sustainable Development Goals (SDGs). The chapters that follow focus on the economic, cultural, and political factors that make up India's unique e-waste processing industry. This systems-level dive into India's e-waste industry demonstrates that there is no *one size fits all* method or law to manage e-waste across various country contexts.

To meet our global goals for human health, sustainability, and economic equality, scholars, industry leaders, and policymakers must attend to the contexts and contributions of the informal processors. Despite the so-called informality of community-based e-waste processors, these small businesses play a critical role in moving the e-waste industry forward. As we will see in this chapter, the critical role that informal processors play in this market is due, in part, to the structure of the e-waste supply chain and the various ways in which e-waste is reused and repurposed. Well before the EPR-based e-waste law became the norm in India, the system already had the foundations of a circular economy.

The EPR framework was first developed for the Swedish context where, unlike India, there was no preexisting waste processing industry upon which to build. It is not surprising that EPR-based e-waste law did not achieve its promised outcomes in India, which already had a thriving preexisting industry.[4] As Chapter 2 will explain, implementation of the e-waste law has resulted in business-as-usual practices that do not lead to better environmental outcomes. While some of the existing practices may look inefficient to an outsider, they are driven by the economic realities of how this industry works in India. Despite these so-called inefficiencies, this system helps keep e-waste out of landfills. By factoring in the advantages of existing market forces that make this system work, we can uncover opportunities to develop a more functional system. In Chapter 3, I delve into the economic factors to derive a deep understanding of the market forces that drive the e-waste industry. Industry leaders, NGOs, and policymakers have long argued that formalization can improve the lives and livelihoods of informal processors through *economies of scale*. Chapter 3 maps the market forces of India's e-waste processing industry such as the nature of e-waste supply, role of international commodity markets, the impacts of changing technologies, and how prices are determined. By delving into greater detail, it shows how this e-waste industry defies *economies of scale* and instead works through *economies of scope*. Informal processors' inherent economic capabilities— low cost, high flexibility, high quality, quick turnaround time—have the potential to maximize economies of scope and align with the ways in which the e-waste industry currently operates on the ground. The existing

---

[4] The EPR-based e-waste law mandated informal processors to formalize and obtain the government approval to process e-waste. The e-waste law also made producers responsible for collection and safe recycling of e-waste.

law that mandates formalization creates obstacles to the effective application of informal processors' inherent capabilities and holds back effective economies of scope.

One key advantage of informal processors' inherent economic capabilities is that they extend the life of electronics products. This strength is built on the foundations of specialized social networks and deep expertise. Chapter 4 will explore the unique cultural context for India's informal processors to lay the groundwork for the development of new policies to leverage local human resources. Generations of cultural and industrial evolution have shaped informal processors' highly-specialized social networks, advancing relationship-driven business practices. This cultural capital yields deep expertise that makes the informal processing system effective. As Chapter 5 explains, the EPR framework that forms the basis of India's e-waste law is rooted in pressures to follow Eurocentric policies. In India, the implementation of this framework has undermined an existing system that already had the potential for culturally-appropriate, sustainable, and economically viable solutions. Chapter 5 proposes e-waste policy futures to illustrate the positive outcomes that can emerge from the trajectories that take local human capital and innovative business strategies into account. Such policies can intentionally leverage the inherent economic capabilities of the informal processors and naturally align with the way India's e-waste industry currently operates on the ground.

This book aligns with global dialogues about e-waste management and broader approaches to formalization policy in the Global South. It examines how we can leverage the strengths of informal e-waste processors to accelerate the transition to a fuller circular economy for India's e-waste industry and offers strategies and framing to support localized policies in other contexts where grassroots "informal" processors already exist.

REDEFINING "INFORMAL" AND "FORMAL" PROCESSORS

Informal is commonly defined as businesses that are not registered.[5] This means that a company is not legally recognized by the government, and whatever business the company does is outside the visibility

---

[5] This is based on commonly accepted definitions of informal and formal (Neuwirth, 2011; Webb et al., 2009).

of the government and not taxed. *Formal* means the company is registered and legally recognized by the government. Whatever business the company does is visible to the government through accounting records and taxed. In the context of the e-waste industry, *informal* means that the processor does not have government approval and that their practices are not being monitored for their potential negative effects on human health and environment. *Formal* means the processor has government approval and monitored for compliance with laws to protect human health and environment.

A recent report estimated India's e-waste industry market size at $1.7 billion and is expected to reach $7.5 billion by 2030 (Avendus, 2023). More than 80% of India's 1.6-million-ton e-waste is recycled by informal processors (Lundgren, 2012; Rajya Dutta & Goel, 2021; Sabha, 2023). Azhar's 200-square-foot scrap shop is a typical example of an informal e-waste processor. This small-scale enterprise processed e-waste items such as mobile phones and televisions. Informal processors engage in rudimentary processing operations with low setup and operating costs for their livelihood. Many of them are intergenerational and community based, relying on tight-knit networks and kinship ties for their supply chains as well as their technical training. They operate in the backyards of their homes or small shops that keep their real estate costs low. Such informal processors are highly-functional networks of specialists that have been handling specialized waste for centuries.

The rationale for formalization is rooted in the reality that informal processors do not follow mainstream safety protocols to protect human health and the environment. This reality is due, in part, to a lack of resources and training. Most informal processors do not invest in the infrastructure and equipment to mitigate the potential negative health and environmental effects of their waste management practices. For example, they do not use any machinery to collect or neutralize the toxic fumes and dust that are generated during e-waste processing. They do not use any protective gear such as masks, goggles, helmets, and gloves during e-waste processing. When the informal processors break down e-waste or recover precious metals such as gold and silver using acids and chemicals without any machinery or protection, they inhale the toxic dust and fumes leading to negative health effects such as itching, cough, shivering of hands, and discomfort in the respiratory tract and throat (Mussarath, 2007). As the owner of Gamma (an informal-turned-formal processor) shared about his days in the informal economy, he was literally *eating* the dust:

> Before, we were eating the dust, dust was not only there, we were eating it [...] we used to clean our nose in water, we used to eat bananas [...] Bananas can take away all the dust from throat and elsewhere.

To ease the discomfort in the throat, he used simple methods such as eating bananas. Some of these toxic dust and fumes are emitted to the surrounding environment. The by-products of e-waste processing such as left-over acids after metal recovery are often disposed unsafely in public places without any treatment or neutralization. These disposal methods lead to negative environmental effects such as ground water pollution that negatively impact human health (Grant et al., 2013; Toxics Link, 2014).

In the introductory example, when Azhar began his formal company Alpha Recycling, he had installed machinery to collect toxic dust, a system to neutralize this toxic dust and other by-products, and the workers used protective gear. There are many informal processors like Azhar who have transitioned to formal and setup companies. For example, Beta, Gamma, Delta, Zeta, and Epsilon are some examples of such informal-turned-formal processors.

Apart from improving the safety aspects, formal processors face logistical advantages and opportunities over informal processors. This is because the e-waste law constricts the operations of informal processors. First, as per the e-waste law, *bulk consumers* are mandated to dispose their e-waste to formal processors only.[6] Informal processors cannot legally buy e-waste from bulk consumers. Second, formal processors can do international business transactions as they are registered companies with government approval for e-waste processing. This means they can apply for the license to export or import products with international business partners. Informal processors cannot do such international business transactions directly, as they are not registered companies with government approval. The third advantage of formalization relates to financing. Banks tend to provide loans to formal processors because,

---

[6] Bulk consumers mean large companies that dispose their e-waste in large quantities. For example, a global technology company and a university disposing 100s or 1000s of laptops, desktops, and other items are bulk consumers.

unlike informal processors, they are registered companies and can typically pledge collateral.

These logistical constraints for informal processors are based on a commonly-accepted definition of informal and formal, as we saw earlier. In the context of India's e-waste industry, this definition is rigid and does not hold fully. Many e-waste processors labeled as *informal* pay their taxes, and yet they are subject to the constraints that formalization policies impose. For example, informal processor Aleph is the owner of Jas—a scrap metal trading company. This is a small shop on the roadside of Bengaluru's scrap recycling neighborhood. Like Azhar's shop, Jas is recognized by the government as has a value added tax (VAT) registration. This means that whenever Aleph buys e-waste from bulk consumers or sells the processed e-waste to downstream players, a tax is levied by the government. Similarly, informal processor Beet owns a scrap metal trading company that has a VAT registration and business transactions taxed by the government. Many such informal processors have VAT registrations and their business activities are visible to the government. These informal processors are popularly referred to as scrap dealers, scrap traders, scrap vendors, scrap merchants, etc. Having a VAT registration means they can be easily identified by the government and are a taxable business. Unfortunately, they are considered as informal by the government, popular media, and industry leaders.

Informal e-waste processors, before the emergence of e-waste law, did not formalize because the terms *informal* and *formal* did not even exist in their vocabulary. The practices later deemed *informal* had long been their *way of living*. It was only during the early 2000s that India's highly skilled specialists that processed e-waste were defined as informal. The owner of Alpha, an informal-turned-formal e-waste processor, explained this situation as follows (edited for readability):

There was no word called 'formal,' everything was 'informal'. Government saw this and said: 'this is informal, that is formal' and divided the subject in two ways. What did they do after dividing? They gave names: 'you are the informal, you be as a formal' [...] Informal is our own source. Government acquired this source. Government watched this, surveyed in all areas and after surveying, they said: 'Due to this, health is affected negatively, environment is damaged, this is a hazardous material'. Nobody knew this before. When we started working [e-waste processing], there was no law above us. When informal, formal were divided and shown, the law was born. Some law had to be made for e-waste, for making law there had to be some reasons, justifications. They made those reasons, 'informal' and 'formal'.

There was no concept of *formal* e-waste processing then. The e-waste industry existed before the e-waste law, without the explicit label of *informal*. For this reason, informal processors are the dominant players in the e-waste processing industry. This is a high-performing industry that is naturally structured to create value from e-waste. The next section explains how India's preexisting e-waste supply chain was set up to create value from e-waste.

E-waste Supply Chain That Naturally Creates Value from E-waste

Unlike a standard supply chain where companies manufacture and distribute their products to consumers, this context is about consumers discarding their used products and the stages they pass through until they are reused and repurposed in useful ways. Well before e-waste law was introduced, India's informal processors played a vital role in the sustainable management of e-waste. Demand for their services in the second-hand markets and the unique strengths and networks within informal processor communities enabled value creation from e-waste through technical innovations and well-developed supply chains. Large corporations relied on them to dispose their e-waste, and much of the population relied on them for second-hand electronics.

The chapters that follow will explore the various dimensions that make India's informal e-waste processors work in spite of the constraints posed by the e-waste law. To better understand the strengths of this market, it is worth first establishing a basic understanding of the many pathways through which informal processors leverage their networks, creative business practices, and deep technical expertise to operate within a highly-specialized industry. As we will see throughout the chapters that follow, these *informal* processes are inextricable from India's broader e-waste processing industry and often necessary for the effective functioning of formalized practices.

Take the case of a Bengaluru-based global technology company, designated as a bulk consumer because it disposes thousands of used laptops, desktops, air conditioners, fans, lighting systems, wiring, and so on. The informal processor Aleph buys these e-waste items and processes them through reuse, repair, and refurbishment. The products that are functional (working condition) and have a high demand in the second-hand market are sold directly to the retailers in the second-hand market. This is simple *reuse*.

Products that cannot be directly reused are *repaired* or *refurbished*. If the products are not functional, but have a demand in the second-hand market, Aleph sells them to another informal processor who specializes in repairing and refurbishing to fix the broken parts and restore the functionality.[7] These are then sold to retailers in the second-hand market.

Products that cannot be reused, repaired, or refurbished are broken down for *commodities trading* and *recycling*. If the products do not have any demand in the second-hand market, Aleph disassembles them into commodities such as plastics, glass, and various types of metals. The plastic casings, wires, steel, circuit boards, etc., are taken out from electronic products, kept separate, and stacked up. The outer cover of wires is removed and copper inside the wire is taken out, kept separate, and stacked up. These disassembled commodities are then sold to specialized commodity traders and recyclers.

The plastic, glass, brass, copper, and other materials are sold to a plastic, glass, brass, or copper traders as appropriate. Each commodity enters a specialized recycling industry that warrants a study of its own. Once the commodities enter their respective industries, they take different forms. Ultimately, these scrap commodities from e-waste find their way into automotive components, household plastic products, and so on. The recycling journeys of plastic and copper, for example, look vastly different in terms of their cultural practices, geographic distribution, supply chains, and other factors. For example, a plastic recycler whom I visited buys scrap plastic from e-waste processors, transforms them into plastic granules, and shapes these granules into buckets, mugs, and other household products. These products are then sold to small-scale retailers across the country. Some of these operations engage both formal and informal processors. While certain commodity traders may be legally registered, for instance,

---

[7] Sometimes informal processors double up as retailers in the second-hand market. For example, I visited Cool's (an informal processor) facility in Bengaluru. The owner of Cool collects e-waste from well-known electronics retailers, repairs them, and sells in his own store located in the city and to other second-hand market dealers far away from the city.

the facility where the plastic recycler melts the hard plastic granules into the form of buckets is not.

In another instance, RML (a commodity recycler in Bengaluru) converts the copper and aluminum disassembled from e-waste to copper metal blocks and aluminum metal blocks and sells them to copper and aluminum smelters, respectively. The smelters transform these scrap commodities into intermediate products and sell them to manufacturers in other industries such as automotives.

Once the e-waste is broken down, metals are recovered from components of the original parts. For example, some informal processors buy circuit boards from Aleph and recover or extract materials such as gold, silver, copper, platinum, and palladium. They may extract metals from the smaller components, such as gold-coated pins embedded in the circuit boards, through chemical treatment. In some cases, processors buy e-waste from bulk consumers and do the metal recovery by themselves. For example, Intro (an informal-turned-formal processor) recovers silver from server boxes, X-ray screens, and keyboard sheets and recovers platinum, silver, and molybdenum from heavy electrical machinery.

In many cases, informal processors recover potentially reusable parts from products that are not functional.[8] For example, new televisions are made from the cathode-ray tubes of desktop computers. The owner of Compost (waste management NGO that works closely with the informal processors) explained about alternate markets for cathode-ray tubes from desktop computers. Cathode-ray tubes are taken out from used computers and tested to check if they are functional. Then, they are refurbished— where the scratches on the screen are cleaned—and sold to television manufacturers in China. Their low-priced televisions are sold in the second-hand markets globally.

The unique strengths of this supply chain with processors such as Aleph, Cool, and specialized commodity recyclers are that nothing is wasted when the markets work. Apart from the Bengaluru-based global tech company selling e-waste, this supply chain works in other cases too, such as a university selling its used washing machines, fans, and lighting from the student hostels, a circuit board manufacturer disposing defective circuit boards, or an importer selling the imported metal scrap. This supply chain existed long before the EPR framework for e-waste law was

---

[8] This is called cannibalization.

planned by India's policymakers and NGOs, and it continues to operate and adapt notwithstanding the barriers that recent policies have imposed.

HISTORICAL CONTEXT OF EPR-BASED E-WASTE LAW

Such a preexisting supply chain was not present in Sweden and other EU countries where EPR was first formulated in the early 1990s, and the implementation of these policies in India shows the pitfalls of stamping Eurocentric policies onto contexts in the Global South. Theoretically, India's e-waste law adopted an approach that had proved successful elsewhere. EPR is intended to reduce the amount of e-waste sent to landfills and to incentivize producers to design greener products. Instead of holding the consumers and the government responsible for waste disposal, it passes the responsibility of waste disposal to the producers (Fishbein, 1998). Under this approach, producers take responsibility for the entire lifecycle of their products, including taking products back from consumers, recycling, and safe disposal (Lifset et al., 2013; Lindhqvist, 2000). For example, Apple is responsible for recycling and disposing of iPads and iPhones after consumers' usage. This concept of passing the responsibility of product disposal to producers was put in place by EU for e-waste in 2002 through the Waste Electrical and Electronic Equipment (WEEE) Directive. The WEEE Directive imposed collection, recovery, and recycling targets for various categories of e-waste on all EU member countries. Slowly, other countries such as Japan, India, and different states in North America began to follow the WEEE Directive as the template for developing their own e-waste policies (Atasu & Wassenhove, 2012).

EPR was an intelligent response to fill gaps created by the lack of preexisting recycling industries in some countries and their government's funding constraints for waste management. It was first introduced by the Swedish Ministry of Environment in 1990 in response to analysis of waste management and recycling systems in nations such as Sweden, Germany, The Netherlands, Austria, Norway, and France (Lindhqvist, 2000). In these countries, though many initiatives were tested to reuse and repurpose the products discarded by consumers, they were not successful due to the lack of an existing recycling industry and high costs incurrent by the government to collect waste from consumers. The high costs of collecting waste were often subsidized by directly charging the consumers or imposing taxes. In these contexts, EPR provided these governments with a mechanism for financing the collection and recycling costs without raising taxes or directly charging consumers.

The fundamental assumptions underlying EPR policy are predicated upon the conditions of the countries for which it was originally designed. EPR was developed to solve a problem that existed due to the lack of a preexisting recycling industry. These conditions are vastly different from contexts such as India, where there was already a preexisting, high-performing e-waste processing industry. As we saw in the earlier section, India's supply chain that created value from e-waste and dominated by the informal processors existed well before the e-waste law. The EPR policies that worked in Scandinavia and elsewhere were inappropriate for the context in India. Unsurprisingly, many e-waste processors in India were skeptical of the new requirements and resisted the constraints imposed on their previously successful business models. More than a decade after introducing the EPR-based e-waste law, informal processors continue to dominate by processing more than 80% of the country's e-waste (Bhargava & Koshy, 2017; Dutta & Goel, 2021; Rajya Sabha, 2023; "Very high non-compliance", 2021). As we will explore in Chapter 2, e-waste processors, producers, and even government officials quickly recognized the challenges that EPR compliance would impose.

In the year 2010, responding to global policy trends and suggestions given by international NGOs, India's government introduced a new law to regulate and rewire the preexisting e-waste processing industry. This was known as *E-waste (Management & Handling) Rules*. By introducing new e-waste law, India's policymakers modeled their legislation on the basic tenets of EPR. Promising better human safety, economic, and environmental outcomes, policymakers urged informal processors to break from their traditional approaches and adopt business and processing practices that have been implemented in developed markets. Policymakers and industry leaders began requiring informal processors to formalize. This meant building facilities with standardized practices in designated areas in order to qualify for business licenses to process e-waste. It became illegal to operate as an informal processor.

"Stick" and "Carrot" That Motivated the Informal Processors to Formalize

The formalization idea was constructed in the early 2000s by large formal processors, policymakers in the government, and NGOs who wished to align India's regulations with policy trends in the EU. The informal processors who formalized did not do it independently by following the e-waste law. They were influenced to formalize by the consultants from Stimulus (an EU-based international development agency), before the e-waste law was officially introduced.

With the support of policymakers and large formal processors, Stimulus' consultants sought to make informal processors aware of the negative health and environmental effects from informal ways of e-waste processing. For example, they visited the informal processors and verbally conveyed the hazards of inhaling acid fumes while doing metal recovery.

Such a direct intervention was not an easy task for Stimulus. Initially, the informal processors were apprehensive and often skeptical. Would Stimulus steal their processing knowledge? Did the Stimulus consultants include officials from income tax department? Consultants from Stimulus secured the trust of informal processors by engaging with them repeatedly over the course of a year. Once the informal processors came to believe that the intention of the consultants was to protect their health and environment, they began to listen. But this safety aspect was not the primary reason that motivated them to formalize. Rather, it was due to the *stick* and *carrot*.

The stick was the forthcoming e-waste law. Many bulk consumers who were disposing e-waste to informal processors began to ask for the approval or license from government.[9] Many bulk consumers started online auctions for e-waste disposal, excluding informal processors from participating in these auctions due to their lack of government approval. Informal processors realized that, if they did not formalize, they would not be able to continue their businesses. On the other hand, if they formalized, they could buy e-waste from bulk consumers and sign long-term deals with them.

---

[9] Partly, this is due to Stimulus influencing the bulk consumers not to dispose e-waste to informal processors. Bulk consumers began to realize the negative health and environmental impacts of informal processing through the popular media and they wanted to protect their brand image.

The carrot was that, if the informal processors formalized, the government or Stimulus would recognize their work, help them to network with large bulk consumers to buy e-waste, and provide better facilities such as a subsidized real estate to do e-waste processing.[10]

Based on this stick and carrot, some of the informal processors decided to become formal and set up formal companies such as Alpha, Beta, Gamma, Delta, Epsilon, Zeta, and many others. Formalization was emotionally draining and financially difficult for the informal processors, and many of these business owners continue to struggle with the impacts of formalization.

THE CHALLENGES OF FORMALIZATION FOR INFORMAL PROCESSORS

Formalization is a lengthy and expensive process that requires significant resources from both informal processors and the government. It requires setting up a company and obtaining government approval, a process that often takes 1–2 years or more. The informal processors begin this process by searching for and securing suitable real estate in industrial areas earmarked by the government. This step is followed by preparing a detailed report of the facility being planned, obtaining initial approval to build the facility, and obtaining final approval to operate the facility after it is built.

The compliance requirements for formalization are misaligned with the processes and practices that work best in India's e-waste industry. Even a requirement as seemingly simple as requiring a table to disassemble e-waste can create barriers to existing methods. Informal processors typically squat on the floor to work on e-waste. This helps them to get their work done quickly, as they can work on a variety of products of different sizes—from mobile phones to televisions or large servers. When workers are required to use a table and chair to disassemble e-waste, it becomes difficult to work quickly on a large variety of e-waste products. As we will see in Chapter 3, the ability to work efficiently on a wide variety of products is a key economic capability to operate in this industry. Formalization increases the time required to process e-waste. In the words of the owner

---

[10] These were the benefits perceived by the informal processors.

of the informal processor Aleph, "*In the informal, one can work fast. But, in the formal we have to work slow.*"

Another major challenge for the informal-turned-formal processors is not being able to access enough e-waste to meet their increased costs of running a formal business. Informal processors rely on low overhead costs to purchase large amounts of e-waste from bulk consumers. The considerable up-front costs of formalization leave them with fewer resources to buy e-waste from bulk consumers, who continue to raise their rates. These constraints force formalized processors to buy, process, and sell less than they were able to prior to undergoing this approval process.

In addition to these economic challenges, the informal-turned-formal processors face multiple government-related challenges. For example, the approval license from the government is given to the physical location of the company. Many aspiring formal processors rent before they can buy so that they can meet the requirement of having a physical location in place. The result is that these businesses often must relocate due to raising rents or requests from landlords to vacate. When this relocation happens, however, the re-application rules require that owners restart the formalization approval process from scratch in their new location. Re-applying for formalization means that they must begin from square one of this lengthy and expensive process, all while losing their initial investment and any government approvals already gained. Many of these processors are then forced to buy and process e-waste in informal ways.

Another government-related challenge is the ad hoc law for license renewal. The government typically provides approval to process e-waste for 3–5 years. After this period, the informal-turned-formal processors need to renew their approval licenses. Recently, the government raised the minimum requirements for facilities to grant these approvals. This means informal-turned-formal processors need to buy additional real estate by expanding their current facility or finding a new location while applying for a license renewal. The owner of Delta explained this challenge as follows:

> At that time, when we made this company, one could get license even if he had only 1,000 sq. ft. land. Now, the new rules say: "you should have land between 4,000 and 8,000 sq. ft., only then we will give the certificate" So, where will we go? If we change the building, we again need to run behind the same head-ache [to get a new license], need to run behind the

pollution control board officers, need to beg and plead: "please give me license."

These challenges strain the informal-turned-formal processors and drain their entrepreneurial energy to do business in compliance with formalization laws.

As we will see in detail in the chapters that follow, formalization has not led to better environmental and economic outcomes in India. Instead, it has constrained enterprises that were thriving at the expense of businesses, communities, and cultural practices. The operations of informal processors before the introduction of formalization show us that there has long been potential in India's e-waste sector for the practices that we know to be essential for a thriving circular economy.

What would be the consequences if the informal-turned-formal processors decide to quit the e-waste industry due to these obstacles and challenges? The manager of Epsilon, who also had an ownership in the business, explained that he would close the formal e-waste company if he were unable to make profits:

> I'm pumping money into e-waste business by taking the profits from my other businesses. Instead of e-waste business, if I had started contract manufacturing for some textile firm, I would have got good profits by now. I will look for 1 more year, if I'm not getting profits, I will close this e-waste business.

Apart from Epsilon's difficulties, there are also instances where some informal-turned-formal processors have returned to their informal businesses. For example, Aleph and Beet returned to their informal roots after quitting their formal processing businesses. If these things happen at scale, a well-functioning supply chain for e-waste would come to a halt, and it is unlikely that the intended environmental outcomes will be achieved. These consequences extend not only to the informal-turned-formal processors themselves, but also to the local schools and communities who rely on the second-hand market routed through the informal processors for access to affordable electronics.

Synthesizing Global Dialogues on Informal Economy and Circular Economy

Global dialogues about e-waste policy continue to assume that the informal economy is a transitory phenomenon which will wither away as cites and nations modernize or frame the informal economy as a *problem* to be dealt with (Guha-Khasnobis et al., 2006). Recent scholarly conversations have begun to turn this traditional thinking upside down (Chen, 2006; Guha-Khasnobis et al., 2006). The informal economy is here to stay and is intrinsically linked with the formal economy. Scholars have noted that we need go beyond the formal-informal dichotomy and reframe formalization policies centered around the informal economy to enable them to unlock their potential. By synthesizing the global dialogues on informal economy, formalization, with the circular economy, this book examines how we can leverage the strengths of informal e-waste processors to unlock their potential and accelerate the transition to a fuller circular economy for e-waste industries.

To make this shift, we first need to understand the strengths that informal processors bring to the table. The effectiveness of these small enterprises to operate within the current constraints of EPR policies speaks to their resilience and innovative business strategies. A closer look at their practices and processes demonstrates that informal e-waste processors have long embodied traits that we associate with a thriving circular economy. To close this chapter, let us consider three ways that India's informal processors have leveraged their inherent capabilities to advance sustainable and profitable e-waste processing. As I hope to show throughout this book, the most salient barriers to a thriving circular economy in India's e-waste industry relate to external stigma, Eurocentric policies, and a lack of holistic understanding of how the industry works.

As we saw earlier in this chapter, India's informal processors play a critical role in processing and repurposing e-waste in useful ways rather than sending it to landfills. Products and parts that can be reused, repaired, or refurbished are sold in the second-hand market. Products that can be disassembled into commodities are being disassembled, and the commodities are being sold to traders and recyclers. The e-waste supply chain is structured to create value from e-waste in three ways: *circling longer, cascaded use, and pure circles*. These are the core tenets of a circular economy (Ellen MacArthur Foundation, 2013).

Let's revisit the earlier examples from the supply chain section. The e-waste disposed by the global tech company or the circuit board manufacturer is not dumped in the landfills or exported to other countries. Rather, they are reused and repurposed in useful ways that extend their life. When Aleph or Cool inspects the products and channels them to the retailers in second-hand market or repair and recover potentially reusable parts, they enable e-waste to *circle longer*. When a discarded keypad is reused first as a second-hand item and later converted into plastic granules for making plastic products such as buckets and mugs, the system enables *cascaded use*. When e-waste is disassembled into commodities that are kept separate, stacked up, and sold to specialized commodity traders and recyclers, the system ensures that one commodity is not contaminated with another commodity and ensures *pure circles*. The uncontaminated scrap increases the quality of metal recovered in the final step.[11]

This working system reduces waste, improves recycling and reuse, and contributes to achieving SGD 12's sub-goal related to waste management.[12] But the advantages of the existing system are not being leveraged to achieve these global goals. This is due to the lack of a systems-level understanding of the existing e-waste industry. The following chapters explain the advantages of the existing system, namely the structure of e-waste industry rooted in *economies of scope* rather than *economies of scale* and the economic capabilities and cultural capital of informal processors rooted in specialized social networks and deep expertise.

If India's government, policymakers, popular media, and other industry leaders understand the economics, cultural capital, and politics in the e-waste industry and reframe the e-waste problem in a holistic manner, they will be well prepared to design or update context-appropriate policies. This book seeks to help dismantle the existing dominant discourse centered around EPR and replace it with one that will support decision makers to leverage the strengths of the mainstream informal processors. When we take a more expansive view of the inherent economic capabilities and cultural capital among India informal e-waste

---

[11] The contamination aspect is explained in Chapter 3.

[12] This existing circular economy directly contributes to UN SDG 12 that aims to achieve sustainable consumption and production in the context of waste management (Schroeder et al., 2019). The specific sub-goal within SDG 12 related to waste management aims to *"substantially reduce waste generation through prevention, reduction, recycling and reuse"* by improving the recycling rate (One Planet Network, n.d.).

processors, it becomes clear that many of the existing challenges within that context can be overcome with better policies. By creating responsive policy frameworks that diverge from the assumptions of EPR, we can strengthen the existing working system to accelerate the transition to a fuller circular economy.

REFERENCES

Avendus. (2023). *Circular economy: Recycling waste to wealth.* https://www.avendus.com/india/reports/61

Atasu, A., & Wassenhove, L. N. (2012). An operations perspective on product take-back legislation for e-waste: Theory, practice, and research needs. *Production and Operations Management, 21*(3), 407–422.

Bhargava, Y., & Koshy, J. (2017, October 11). 225 companies pulled up for e-waste rules violations. *The Hindu.* https://www.thehindu.com/news/national/225-companies-pulled-up-for-e-waste-rules-violationsdeck-they-are-supposed-to-collect-and-scientifically-recycle-it/article19840729.ece

Chen, M. (2006). Rethinking the informal economy: Linkages with the formal economy and the formal regulatory environment. In G. Basudeb, R. Kanbur, & E. Ostrom (Eds.), *Linking the formal and informal economy: Concepts and policies* (pp. 75–92). Oxford University Press.

Dutta, D., & Goel, S. (2021). Understanding the gap between formal and informal e-waste recycling facilities in India. *Waste Management, 125*, 163–171.

Ellen MacArthur Foundation. (2013). *Towards the circular economy Vol. 1: An economic and business rationale for an accelerated transition.* https://ellenmacarthurfoundation.org/towards-the-circular-economy-vol-1-an-economic-and-business-rationale-for-an

Fishbein, B. K. (1998). EPR: What does it mean? Where is it headed? *Pollution Prevention Review, 8*(4), 43–55.

Grant, K., Goldizen, F. C., Sly, P. D., Brune, M. N., Neira, M., van den Berg, M., & Norman, R. E. (2013). Health consequences of exposure to e-waste: A systematic review. *The Lancet Global Health, 1*(6), e350–e361.

Guha-Khasnobis, B., Kanbur, R., & Ostrom, E. (2006). Beyond formality and informality. In G. Basudeb, R. Kanbur, & E. Ostrom (Eds.), *Linking the formal and informal economy: Concepts and policies* (pp. 1–18). Oxford University Press.

Lifset, R., Atasu, A., & Tojo, N. (2013). Extended producer responsibility: National, international, and practical perspectives. *Journal of Industrial Ecology, 17*(2), 162–166.

Lindhqvist, T. (2000). *Extended producer responsibility in cleaner production* (Doctoral dissertation). Lund University. Lund University Research Portal. https://www.iiiee.lu.se/thomas-lindhqvist/publication/e43c538b-edb3-4912-9f7a-0b241e84262f

Lundgren, K. (2012). *The global impact of e-waste: Addressing the challenge.* International Labour Office. https://www.ilo.org/sector/Resources/public ations/WCMS_196105/lang--en/index.htm

Mussarath, M. (2007). *Study on e-waste recycling processes leading to occupational health hazard in Bangalore* (Master's thesis). Bangalore University.

Neuwirth, R. (2011). *Stealth of nations: The global rise of the informal economy.* Random House.

One Planet Network. (n.d.). *Target 12.5 waste reduction.* SDG 12 Hub. Retrieved January 26, 2023, from https://sdg12hub.org/sdg-12-hub/see-progress-on-sdg-12-by-target/125-reduce-waste-rrr

Rajya Sabha. (2023). *Management of e-waste.* https://pqars.nic.in/annex/259/AS5.pdf

Schroeder, P., Anggraeni, K., & Weber, U. (2019). The relevance of circular economy practices to the sustainable development goals. *Journal of Industrial Ecology, 23*(1), 77–95.

Thierry, M., Salomon, M., Van Nunen, J., & Van Wassenhove, L. (1995). Strategic issues in product recovery management. *California Management Review, 37*(2), 114–136.

Toxics Link. (2014). *Impact of e-waste recycling on water and soil.* https://ipen.org/sites/default/files/documents/Impact-of-E-waste-recycling-on-Soil-and-Water.pdf

Very high non-compliance of e-waste rules at 72 collection centres in Delhi. (2021, July 4). *The New Indian Express.* https://www.newindianexpress.com/cities/delhi/2021/jul/04/very-high-non-compliance-of-e-waste-rules-at-72-collection-centres-in-delhi-pollution-control-board-2325386.html

Webb, J. W., Tihanyi, L., Ireland, R. D., & Sirmon, D. G. (2009). You say illegal, I say legitimate: Entrepreneurship in the informal economy. *Academy of Management Review, 34*(3), 492–510.

CHAPTER 2

"Business as Usual": Systemic Barriers to Achieving Sustainability in India's E-waste Processing Industry

Abstract India's current e-waste law is predicated upon the assumption that formalization is better for human health and the environment. This chapter will examine that claim in context by surfacing its underlying assumptions and exploring whether India's formalization policies have resulted in improved health or sustainability outcomes. The argument in favor of formalization depends on the assumption that the law is followed and enforced. In reality, the systematic workarounds, reciprocal strategies, and intrinsic challenges of the system have resulted in business as usual. By factoring in the advantages of market forces that makes this system work, we can uncover opportunities to develop a more functional system.

Keywords E-waste legislation · Compliance · Enforcement · National recycling rate · Sustainable development goals · Informal recycling · Extended producer responsibility

© The Author(s), under exclusive license to Springer Nature Switzerland AG 2023
T. S. Krishnan, *Untapped Knowledge in India's E-Waste Industry*, Palgrave Advances in the Economics of Innovation and Technology, https://doi.org/10.1007/978-3-031-50296-5_2

INTRODUCTION

As we saw in Chapter 1, India had a thriving e-waste processing industry before the introduction of laws mandating formalization. In this chapter, I examine the assumptions underlying India's formalization policy and assess whether this policy has resulted in improved sustainability outcomes. In particular, I break down the flaws inherent in evaluating sustainability outcomes based on the national recycling rate. As this chapter shows, formalization policies such as Extended Producer Responsibility (EPR) legislation and the adoption of the national recycling rate metric have not been effective mechanisms to achieve sustainability in India's e-waste context. The economic incentives and logistical constraints in India's e-waste industry have led to a broken feedback loop. Both formal and informal processors develop strategies for apparent compliance without advancing the intended environmental and human health benefits of these policies.

As we saw in Chapter 1, the concept of the formal processor in e-waste industry is a social construct of the 2000s. In recent decades, India has adopted the e-waste laws for formalization based on Eurocentric EPR frameworks. Since 2011, this legislation has mandated that all e-waste processors must achieve governmental approval by implementing standard processes and equipment. Many policymakers and industry experts assume that this EPR-based e-waste legislation will reduce waste through reuse, repair, and recycling and protect human health and safety. This assumption was strengthened when the United Nations Environmental Program (UNEP) introduced the national recycling rate as a core indicator to measure the effectiveness of e-waste management in every country (One Planet Network, n.d.). In simple terms, the national recycling rate is the quantity of e-waste that a country formally collects and recycles divided by the total quantity of e-waste that it generates (Forti et al., 2020; One Planet Network, n.d.).

The national recycling rate is used to achieve specific targets of United Nations SDG 12. The nature of India's e-waste processing context leaves both sides of this equation open to ambiguity and manipulation. On paper, India has reported e-waste national recycling rates averaging 1%, positioning the country's recycling industry on a lower scale when compared to the 60% recycling rates documented in Sweden and Switzerland (Forti et al., 2020). Since it is only possible to measure the quantity of e-waste being documented, however, India's metrics are skewed by the exclusion of informal processors. Its comparatively low number does not reflect the realities on the ground. The systematic barriers and opportunities within India's e-waste processing industry have promoted the use of workarounds and maintained business as usual. This chapter maps the systemic challenges created by existing formalization policy and the strategies that both informal and formal processors implement to address those challenges.

SYSTEMATIC WORKAROUNDS PRACTICED BY E-WASTE PROCESSORS

Proponents of formalization make the case that formal processors yield better health and environmental outcomes than their informal counterparts. On paper, the e-waste law in place requires that formal processors take steps toward sustainability such as wearing personal protective gear, disassembling e-waste on a table with dust collection machinery installed, and giving away toxic unprocessed portions for scientific landfilling. This argument in favor of formalization depends, however, on the assumption that the legislation is followed and enforced. In reality, India's e-waste law has resulted in practices that are difficult or impossible to follow.

This reality became clear to me when, during my fieldwork, I visited Gamma's[1] facility located on the outskirts of Bengaluru. I arrived to find an opaque metal gate blocking the view of a closed facility. I called the owner of Gamma from the roadside and waited as an employee checked that I had arrived alone and was neither a government official nor a journalist. Following these precautions, I was permitted to enter through a barely open gate that was quickly locked behind me. I entered the facility and saw three employees disassembling printed circuit boards. They squatted on the floor, taking e-waste apart with bare hands using hammers and pliers. Their helmets, gloves, goggles, and other personal protective gear lay on a nearby table. The unused table was also outfitted with the dust collection machinery that, according to formalization laws, is required for e-waste disassembly. As the owner of Gamma explained, the electricity required to use the dust collection machinery is expensive enough to erode the processor's profit margins. Employees know to use designated gear and equipment for scheduled government inspections.

---

[1] Gamma is an informal-turned-formal processor.

Another informal-turned-formal processor in Bengaluru, which I will call Alpha, echoed the same constraints and workarounds implemented at the Gamma facility. The owner of Alpha also spoke to the constraints of government-mandated safe disposal as well as the practices that this facility uses to dispose of e-waste more affordably and maintain the appearance of compliance:

> We are not getting profits due these various costs. We have no option other than to flout the guidelines. Who cares for the environment, if you cannot make profits. For example, we let the unprocessed hazardous waste down the municipal drainage because we must pay money to TSDF[2] to take this.

Formal processors are required to give away the toxic unprocessed portions of their e-waste for safe disposal (also known as scientific landfilling). To avoid the considerable cost of this step, Alpha dumps the facility's toxic waste along with municipal waste or in open areas.

In another instance, I accompanied the owner of informal processing company Aleph to a bulk consumer—a global technology company—to collect e-waste. The owner of Aleph had borrowed Alpha's license. Alpha, which has a formal facility approved by the government, trades its license with informal processors such as Aleph in return for a commission. The owner of Aleph used this borrowed license to collect e-waste from the global technology company.

Workarounds like these are practiced widely by both formal and informal processors as a necessary economic imperative. It has become commonplace for businesses that are formal on paper to set up showroom factories or buy and sell licenses.

Many formal processors sell e-waste to informal processors to move business forward. While I was interviewing the owner of Beet, a small roadside informal e-waste processor in Bengaluru, a large truck arrived loaded with monitors, CPUs, keyboards, and servers. The owner said that they had purchased this e-waste from a reputed formal processor based in Bengaluru.

The mechanisms that formal and informal processors adopt to advance *business as usual* impact not only local operations, but also national and

---

[2] TSDF means Treatment, Storage, and Disposal Facility. They are companies that provide safe waste disposal services.

international e-waste systems. On paper, a global technology company may give its e-waste to a formal processor as per the e-waste law. But this e-waste eventually gets redirected to informal processors, leading to negative outcomes for human health and the environment and undercutting policy goals. Many policymakers, industry leaders, and other stakeholders blame these negative impacts on informal processors and the Indian government's failure to enforce compliance. The true root cause is more nuanced: Existing legislation makes it impossible for formal and informal processors to operate their businesses safely and effectively. In the current system of India's e-waste industry, businesses cannot succeed economically without workarounds. The owner of RTH, a large formal processor operating in Maharashtra, remarked:

> We don't want to adopt corrupt practices. But, for survival, we need to adopt them. We fall in line with them. I wish, we do not go in that path. But, if I don't go in that path, I will not survive. I need to shut down my business.

A formal processor decides to sell some portion of the collected e-waste to an informal processor to satisfy one of the three economic necessities:

1. A formal processor needs to make revenue by processing the collected e-waste *as quickly as possible*. If the quantity of e-waste is low or has low revenue realizing potential, rather than spending a week on processing this e-waste by hiring contract employees at daily rates, the formal processor prefers to sell to an informal processor. This helps the formal processor to *quickly* make revenue from the collected e-waste and utilize the time efficiently to source high quantity or high value e-waste. If the quantity of collected e-waste was high or has a high revenue realizing potential, the formal processor will not sell this to an informal processor.
2. A formal processor sells e-waste to an informal processor due to a lack of capacity or skilled employees at the facility. In many cases, a mismatch between volume disposed and available capacity occurs when bulk consumers need to dispose a large amount of e-waste at a time. For example, a bulk consumer who is changing office locations and looking to dispose 300 tons of e-waste may insist that the processor purchases the full amount, even if the processor is only equipped to take a third of that amount. The formal processor

cannot afford to turn down this volume and purchases the entire amount. The excess is sold to an informal processor.

3. A formal processor sells e-waste to an informal processor to avoid high inter-state transportation costs. These are the total costs, including taxes, incurred to transport e-waste from one state to another by road. This is particularly relevant when the company disposing e-waste is located in one state and the formal processor buying e-waste is located in another state. In India, the total cost of transporting goods by road accounts for 16% of the gross domestic product, making it expensive when compared to developed countries that have often low single-digit percentages ("Govt aims to", 2023). A formal processor located in Karnataka buying e-waste from a company located in the neighboring state Kerala would face high transportation costs. If the e-waste purchased cannot generate high enough revenues to offset the high transportation costs, it is more economical for the formal processor to sell part or all of the e-waste to an informal processor within Kerala.

Reciprocal Strategies in the E-waste Industry

Formal processors rely on informal processors and their extended network of enterprises for workarounds and to move their businesses forward. As per the e-waste law, bulk consumers are legally required to engage formal processors to dispose of their e-waste. The formal processors invite informal processors to accompany them while they visit bulk consumers such as global technology companies and electronics manufacturing companies to inspect the e-waste. As Chapters 3 and 4 explain in greater depth, the tacit knowledge shared by many informal processors often makes them better prepared than formal processors to assess diverse types of e-waste on site and determine the optimal purchase price based on visual inspection. They can quickly assess the disposed laptops, refrigerators, monitors, and many such e-waste products in terms of what material can be potentially reused, repaired, or recycled.

Informal processors bring critical expertise as well as connections without which India's e-waste industry could not function. In a joint interview with the business development manager of a large formal processor and the owner of the informal processor Aleph, both spoke about their joint business pursuits. As we saw earlier, Aleph often used the license of informal-turned-formal processor Alpha to buy and process e-waste. In this case, however, the owner of Aleph accompanied the business development manager to warehouses of large bulk consumers for assessment and pricing. The formal processor, who employs the business development manager, gave a commission to the informal processor Aleph for this service.

There is a high demand from formal processors for connections to skilled informal processors who can help them overcome roadblocks in the market. In order to thrive, formal processors must build the trust relationships required to do business with informal processors. During my fieldwork, I found that many newly-established formal processors were struggling to establish relationships with informal processors. Many newly-established formal processors asked me to connect them with informal processors to support them with assessment and pricing. Many formal processors have sought to hire informal processors to do processing work in their formal facilities. In an interview, one formal processor based in Bengaluru indicated that they hire informal processors on a per-day basis to come to their facilities to break down the e-waste.

Apart from relying on the informal processors, the formal processors also rely on the wider informal sector. This is an extended network of specialized enterprises comprising informal commodity traders and recyclers who convert the commodities disassembled by processors into useful products for other industries. Formal processors sell disassembled e-waste commodities such as plastics, copper, and brass to informal commodity traders. As we saw earlier, the formal processor RTH did workarounds as a necessary economic imperative. RTH often sells copper and other metal items recovered from e-waste to specialized informal commodity traders. This is because informal commodity traders can pay a higher price, purchase immediately vis-à-vis formal traders, and do not impose a minimum quantity ceiling. These transactions are not taxed and not accounted for in the company records.

Despite being denied governmental legitimacy, the informal processors and the wider informal economy offer these critical workarounds, and formal processors cannot function without their collaboration. These interdependencies indicate that the informal economy is not a *fringe alternative* or a *residual* to the formal economy, but rather the *mainstream* or *lifeblood* of e-waste industry as a whole. The informal economy's mainstream dominant mode is illuminated from a global IT hardware manufacturer's experience of complying with the e-waste legislation. In response to the EPR-based e-waste law, the global manufacturer organized a pilot program in major cities to collect e-waste from retail consumers (members of the public). The pilot program offered monetary incentives to retail consumers to return their used laptops, desktops, monitors, and other IT products of any brand by visiting the exclusive stores of this manufacturer. Though many consumers turned up, very few returned their products for the price offered. This was because in the same city where the stores were located, the informal processors offered a higher price than what the manufacturer offered. Choosing the more lucrative option, the consumers sold their used products to the informal processors. When the manufacturer increased their offer price, the informal economy also increased their price. As the price war escalated, it became impossible for this global manufacturer to compete with India's informal processors.

Systematic Workarounds Practiced by Bulk Consumers and Government

It is not only the formal or informal-turned-formal processors that use these workarounds. Stakeholders at every level of the industry have opportunities to game the system, and many of them do exactly that. Bulk consumers are legally required by the e-waste law to dispose their entire e-waste to formal processors, but many of them instead dispose the high economic value items to informal processors and the low economic value items to formal processors. This is because informal processors pay a higher price than formal processors for the same materials. Informal processors can afford to pay a premium for e-waste due to low operating costs and superior tacit knowledge of e-waste processing. For example, a global technology company may sell minor items with low value such as wires, lamps, bulbs, and worn-out keyboards to formal processors and larger high value items such as desktops, air conditioners, refrigerators, and telecom cables to informal processors. In this way, they can meet government legislation on paper and earn higher revenues from selling e-waste.

Many bulk consumers dispose of e-waste by conducting an auction and selecting the formal processor who quotes the highest amount for the e-waste. Bulk consumers will ask for a copy of the formal processor's license during the initial stage when formal processors submit their quotes. Sometimes, these auctions are often run in unfair ways. For example, some bulk consumers will create an illegal copy of a processor's license and share it with an alternative personal contact who can then purchase the bulk consumer's e-waste under that forged identification. Official records are fudged accordingly. This personal contact could be an informal processor or a formal processor who pays a commission in return for this favor. For example, owner of AMY (a formal processor in Maharashtra) said:

> Sometimes, companies also do cheating. They will take a copy of license from us when we give our quote. Then, they give this license to their contact and sell them the material. We come to know of this after they have given the material. This is illegal. Companies should not be doing this. The Director of the company is not aware of all this. These are done by the purchase officer of that company. He will get some commission for this. That is why he is doing these things.

In addition to these workarounds, many bulk consumers and formal processors game the system by providing kickbacks to the government officials responsible for monitoring their compliance with e-waste law. Bulk consumers are mandated by these officials to dispose e-waste to a specific formal processor at a specific price. Formal processors often pay a commission to government officials for recommending their names to the bulk consumer, and bulk consumers often pay government officials to overlook violations of environmental laws. For example, the business development manager of a formal processor narrated the experience of a bulk consumer that disposed of waste batteries in the company's open backyard area. A government official witnessed this activity during a routine inspection and said: "*You are not supposed to store used batteries in this way. I will issue a notice and you have to pay penalty.*" The official demanded a ₹10,000 kickback, and the company was obliged to pay to avoid a higher penalty if the official were to legally report the incident. In this case, the batteries found were not particularly valuable e-waste materials. If there were large valuable e-waste stored in the open backyard area, then the official might have ordered to dispose of the e-waste to a specific processor at a specific price. If the e-waste were worth ₹10,000, the official would ask to sell at ₹2,000. This helps the processor to buy at a price lower than market value and sell the outputs at a higher price. The processor pays a commission from this transaction to the official. The business development manager explained this as follows:

Manager has no other option. Because, otherwise a legal notice would be issued and Manager may lose his job in the company. On the other hand, companies also find this option better. Else, they will have to pay fine, go to court and all that. So, if companies bribe the inspector [official] or does something according to his wish, he will not issue any notice. So, many inspectors in our country have become crorepatis [millionaires].

Thus, in an industry that relies on workarounds to function, many of the regulations put in place to protect human health and the environment have fallen by the wayside. At the surface level, the government officials do not seem to enforce the e-waste law. While some of these corrupt practices are driven by money, others result from genuine challenges that make it difficult for officials to enforce e-waste law and related regulations.

SYSTEMIC CHALLENGES FACED BY THE GOVERNMENT

Like e-waste processors and consumers, government officials find workarounds to navigate the complicated politics of India's e-waste industry. Government compliance practices perpetuate *business as usual* through flawed and inconsistent enforcement mechanisms. For example, government departments such as the pollution control boards face their own constraints such as labor force shortages and information asymmetries to enforce and monitor regulations. There is widespread public opinion that India's government neither monitors nor enforces the EPR-based e-waste law and that, as a result, producer brands are not complying. While this is often the case, the government faces labor shortages and information asymmetry that hinder monitoring and enforcing the e-waste law.

Labor shortages, for example, present a significant barrier to enforcement of e-waste law because there are simply not enough boots on the ground. As a senior government officer explained in an interview, the state pollution control boards lack resources to monitor and enforce any environmental law due to acute employee shortages. Each compliance officer is responsible for learning, monitoring, and enforcing 15+ environmental laws (including e-waste) across a geographical territory that may have as many as 2,000 companies. The law requires that each company will be evaluated in person on an annual basis or periodically within a year. It is practically infeasible for this officer to physically inspect every company. The EPR-based e-waste law is only one of the environmental laws within the officer's purview. The officer is also responsible for monitoring and enforcing other environmental laws created for batteries, water, air, municipal solid waste, biomedical waste, hazardous waste, plastics, and related regulations.

Typically, it takes a full day for an officer to inspect one company to assess if the company is complying with one law. A single inspection involves traveling to the site, inspecting the specific systems and processes, collecting evidence, and verifying company records. One officer can only handle 100–200 companies per year. So, it is beyond the officer's capacity to monitor all companies for all environmental laws. Every year, the number of companies increases by 10%, and new laws and amendments are created by the Environmental Ministry. But there has been no proportionate increase in the number of officers, and the increased workload is handled by the single officer.

Government labor shortages extend beyond the officials monitoring sites on the ground. Typically, when the officer inspects a company, a physical sample is collected and sent to the government laboratory to test whether the company is following the standards laid out in the law. For example, if the quantity of toxic substances in electronics—such as lead in printed circuit boards—is above a certain threshold, that material is classified as hazardous waste and not e-waste. The hazardous waste cannot be processed to recover metals to prevent the potential negative impact on humans and must be safely disposed of by the processor. The level of such toxic substances can be ascertained only by analyzing a physical sample of the material in the laboratory. The laboratory has one lab specialist who can analyze a maximum of 16 samples per month; however, the lab specialist receives 700 samples per month. The laboratory does not have enough specialists to analyze the samples in need of testing.

As a senior government officer suggested to me, the Environmental Ministry continues to create new laws and amendments and push them to the pollution control boards without understanding the challenges on the ground. Labor shortages throughout the system are only some of the key barriers to translating policy into practice.

Information asymmetry, where the processors have more accurate information on e-waste processing than the government, presents another significant barrier to enforcement of e-waste law. A government officer explained that their reliance on formal processors' self-reporting poses roadblock to enforcing and monitoring e-waste law. Even if the government officer had the resources to inspect every formal e-waste processor, there is no way the officer could verify whether the numbers reported by formal processors are genuine. The government officer narrated his experience of inspecting a formal processor's facility. The documentation revealed that the formal processor purchased only 70 tons e-waste in that specific year, even though the installed capacity was 300 tons. The government officer must believe the numbers presented in the accounting documentation and other records. As it became clear later, this formal processor had purchased more than 70 tons and sold them to retailers in the second-hand market due to the higher profit margins. There are no documents or bills recorded for these transactions. The formal processor confided this reality to the government officer after the officer built a trust relationship.

Another challenge is to scientifically ascertain and verify the records shown by the formal processor. When the officer visits for inspection, the formal processor shows a record of the quantity and type of e-waste (such as mainframes, personal computers, etc.) purchased as well as how much quantity of metals were recovered from them. The officer, who is not an e-waste specialist, must believe what they are told because there are no established standards to verify against. There are no established scientific standards that explain the quantity and type of metals that can be recovered from the various mainframes and personal computers.

In rare cases when government officials overcome these obstacles and successfully demonstrate that formal processors, bulk consumers, or producer brands are not complying with current policies, designated penalties are extremely difficult to enforce. As one senior officer explained, closing down a bulk consumer, processor, or producer is also labor intensive. The officer must collect physical samples from the company, prepare a detailed analysis report, and, in cases where companies are releasing toxic substances, submit the report directly to the local court. If the company is not discharging a poisonous substance with an immediate threat to life, the official can only send warning letters to the company before filing a case in the local court. The company can choose to respond to these warnings by modifying their process and adhering to the legislation. Generally, however, many large companies use political influence to avoid legal consequences. They use political connections to apply pressure to the pollution control boards and their officials not to act against the company.

This is because formal processors, bulk consumers, and producer brands also generate higher tax revenues for the government and employment for the people. For instance, a global technology company providing jobs to more than 10,000 people and paying more than 6-figures tax cannot be closed down only because they are not complying with the e-waste law. Large influential companies in other sectors continue to pollute, and pollution control boards are unable to take any action. For example, the senior government officer mentioned the case of a large oil refining company—this company does not have an effluent treatment plant, and, if any officer decides to take an action against this company, the officer will be transferred to another location. The senior government officer remarked that when such large influential companies in other sectors get away unpunished, it is not reasonable to penalize small-scale relatively non-polluting informal or informal-turned-formal e-waste processors who do this business for a living.

Looking at Inefficiencies Through Another Lens

The systematic workarounds, reciprocal strategies, and systemic challenges of the government system ensure that the e-waste national recycling rate metric does not reflect the reality on the ground. As this chapter has argued, existing policies perpetuate the need for workarounds in India's e-waste sector. Critiques of this system often emphasize inefficiencies and corruption, frequently blaming informal processors for these systemic problems. In practice, however, e-waste processors play a vital role in supporting the sector as a whole through niche expertise, effective networks, and models of operations that are well suited to the e-waste industry. They drive India's vibrant second-hand market and process more than 80% of the country's e-waste (Avendus, 2023; Rajya Sabha, 2023). For example, owner of Delta narrated his experience of purchasing defective mobile phones from a global phone manufacturer. Delta repairs and refurbishes these phones and sells them to dealers in the second-hand electronics markets in Bengaluru, Chennai, and Delhi. These dealers sell the phones at affordable prices that are much lower than the price of the original. E-waste processors such as Delta and these electronics dealers, often classified as unorganized or grey or informal, are a core part of India's $10 billion second-hand electronics market (Gutgutia, 2022).

When we set aside the divide between formal and informal processors and instead examine the strengths of India's e-waste industry, pathways emerge for a safer, more sustainable, and more economically viable system. Even with so-called inefficiencies, which are often driven by economic imperatives, informal processors play a key role in keeping e-waste out of landfills. Even though bulk consumers dispose to informal processors, formal processors sell to informal processors, government officials receive kickbacks, and so on, ultimately the e-waste gets processed and repurposed in useful ways. Products in the e-waste stream that can be reused, repaired, or refurbished are sold in the second-hand market, products that can be disassembled into commodities are being disassembled, and the commodities are being sold to traders and recyclers, and other buyers. In an article that appeared in the journal *Nature*, Walter Stahel—who laid the foundations for circular economy research—explained that this ability to reuse, repair, refurbish, and recycle materials using the market forces is the core principle of establishing a circular economy (Stahel, 2016). As we saw in Chapter 1, India's e-waste industry is already structured to create value from e-waste by *circling longer, cascaded use, and pure circles*—three core elements of a circular economy (Ellen MacArthur Foundation, 2013). Creating such a working system is extremely hard—it is labor intensive and expensive (Stahel, 2016). By factoring in the advantages of existing market forces that makes this system work, we can uncover opportunities to develop a more functional system. In Chapter 3, I delve into the economic factors to derive a deep understanding of the market forces that drive the e-waste industry.

REFERENCES

Avendus. (2023). *Circular economy: Recycling waste to wealth.* https://www.avendus.com/india/reports/61

Ellen MacArthur Foundation. (2013). *Towards the circular economy Vol. 1: An economic and business rationale for an accelerated transition.* https://ellenmacarthurfoundation.org/towards-the-circular-economy-vol-1-an-economic-and-business-rationale-for-an

Forti, V., Baldé, C. P., Kuehr, R., & Bel, G. (2020). *The global e-waste monitor 2020.* United Nations University, International Telecommunication Union, and International Solid Waste Association (ISWA). https://ewastemonitor.info/wp-content/uploads/2020/11/GEM_2020_def_july1_low.pdf

Govt aims to bring down logistics cost to 9% by 2024. (2023, March 28). *Mint*. https://www.livemint.com/news/india/govt-aims-to-bring-down-logistics-cost-to-9-by-2024-gadkari-11680011038160.html

Gutgutia, M. (2022, May 2). The $10Bn used smartphone market to watch out for. *Redsights*. https://redseer.com/newsletters/the-10bn-used-smartphone-market-to-watch-out-for/

One Planet Network. (n.d.). *Target 12.5 waste reduction*. SDG 12 Hub. Retrieved January 26, 2023, from https://sdg12hub.org/sdg-12-hub/see-progress-on-sdg-12-by-target/125-reduce-waste-rrr

Rajya Sabha. (2023). *Management of e-waste*. https://pqars.nic.in/annex/259/AS5.pdf

Stahel, W. R. (2016). Circular economy. *Nature, 531*, 435–438.

CHAPTER 3

The Power of Markets: Economics of India's E-waste Processing Industry

Abstract This chapter challenges the widespread assumption that formalization can improve the lives and livelihoods of informal processors through *economies of scale*. This assumption is rooted in a biased understanding that ignores the existing strengths and future potential of India's informal processors. This chapter maps the economic aspects of India's e-waste processing industry and shows how e-waste industry defies *economies of scale* and rather works on *economies of scope*. Informal processors' economic capabilities—low cost, high flexibility, high quality, quick turnaround time—are such that they maximize economies of scope and align with the ways in which the e-waste industry operates. The existing e-waste formalization policy obstructs the inherent capabilities of informal processors that naturally maximize economies of scope. If India were to redesign its policies to work in collaboration with informal processors, both sides would stand to reap massive economic rewards.

Keywords E-waste recycling · Business models · Recycling economics · Informal recycling · Economies of scale · Formalization

© The Author(s), under exclusive license to Springer Nature Switzerland AG 2023
T. S. Krishnan, *Untapped Knowledge in India's E-Waste Industry*, Palgrave Advances in the Economics of Innovation and Technology, https://doi.org/10.1007/978-3-031-50296-5_3

INTRODUCTION

As we saw in Chapter 2, current policies in India's e-waste industry perpetuate a system in which both formal and informal processors rely on workarounds and reciprocal strategies to operate their businesses. These widespread barriers notwithstanding, informal processors play a critical role in keeping e-waste out of landfills and facilitating benefits associated with thriving markets for second-hand electronics and commodities. As we saw in Chapter 1, India's e-waste industry is already structured to create value from e-waste. Creating such a working system from the ground up is extremely hard. It is labor intensive—requires people with skills to reuse, repair, refurbish, disassemble, and recycle. It is expensive—governments often need to fund the setting up of a working system through taxation schemes (Stahel, 2016). By factoring in the advantages of existing market forces that makes this system work, we can uncover opportunities to develop a more functional system. This chapter delves into the economic factors on the ground in India to derive a deep understanding of the market forces that drive its e-waste industry.

Industry leaders, NGOs, and policymakers have long argued that formalization can improve the lives and livelihoods of informal processors through *economies of scale*. This chapter challenges the widespread assumption that formal processors are more economically viable than their informal counterparts. This assumption is rooted in a biased understanding that ignores the existing strengths and future potential of India's informal processors. This chapter maps the economic aspects of India's e-waste processing industry such as market forces, the impacts of changing technologies, and how prices are determined. Through this mapping, it shows how e-waste industry defies *economies of scale* and rather works on *economies of scope*. Informal processors' inherent capabilities include low cost, high flexibility, high quality, and quick turnaround time. These capabilities are poised to maximize economies of scope because, in large part, they are aligned with the way in which the e-waste industry operates. When informal processors formalize, as we have seen, new regulations create roadblocks that hurt informal processors instead of helping them achieve stated objectives for economic gain, human health, and environmental sustainability.

If India were to redesign its policies to work in collaboration with informal processors, both sides would stand to reap massive economic rewards. To fully realize these potential economic benefits, India must recognize and work in cooperation with its robust local economies and, in particular, the human resources who have developed highly functioning systems in the face of systemic adversity.

FORMALIZATION ASSUMES ECONOMIES OF SCALE

In Chapter 1, I explained the formalization process, challenges introduced by formalization, and how some informal-turned-formal processors returned to their informal businesses. While a few informal processors who formalize do reap economic benefits, there are many other factors at play. Informal processors who do thrive after formalizing achieve this success not because of formalization itself, but because of the preexisting business partnerships and strategies that facilitate collaboration across formal and informal contexts. Take the case of Delta, an informal-turned-formal e-waste processor. The owner of Delta said that he was able to purchase more e-waste and make more profits after becoming formal. A short tour inside Delta's facility corroborated this. Delta's processing facility in Bengaluru was a two-story building with two office rooms and a closed-circuit television with four cameras: one covering the office and three covering the shop floor. Delta had two leading International Organization for Standardization (ISO) certifications hung on the wall of the main office room. The shop floor was filled with e-waste, and employees were busy in processing operations.

Delta's success is not because of formalization itself, but because Delta leveraged prior partnerships and changes in the external business environment. Before becoming formal, Delta has been doing business with a large telecom company. When Delta became formal, there was a shift from 2G to 3G in the telecom industry. This forced all of the telecom companies to replace their towers and other equipment with 3G compatible systems. The preexisting partnership facilitated Delta to purchase the India-wide e-waste consisting of 2G compatible towers, servers, and other materials from this large telecom company. This single large deal was enough for Delta to become profitable without the need to purchase any e-waste for six months. Later, even though Delta faced challenges to purchase e-waste, the owner leveraged his professional connections with other stakeholders such as bulk consumers, the wider informal economy such as second-hand market dealers, and commodity traders to keep the business moving forward. Formalization in itself did not enable Delta to purchase *more* e-waste from *more* bulk consumers or to produce *more* outputs by processing *more* e-waste at lower costs. This shows that even Delta could not *scale* after becoming formal.

The fundamental assumption made by the industry leaders, NGOs, and policymakers is that formalization would improve the business of informal processors through economies of scale. The assumption is that when an informal processor becomes formal, *more* e-waste can be collected from *more* bulk consumers, and this will help the processor to process *more* e-waste and produce *more* outputs at lower unit costs. As we saw in Chapter 1, when the formalization process began during the mid-2000s, many bulk consumers who had previously been disposing e-waste to informal processors began asking for government approval due to the forthcoming e-waste law. Many bulk consumers also began online auctions for e-waste disposal, and it was impossible for informal processors to participate in these auctions due to the lack of government approval. Seeing these trends, many informal processors believed that formalization would enable them to grow through *economies of scale*.

Scale is often equated with the idea of a large, centralized, factory-style assembly line where repeatable standardized process is used to produce a large quantity of similar products at lower unit costs. However, this assumption about *economies of scale* is based on an incomplete understanding of the economics of e-waste processing industry. For processors to realize economies of scale and operate like a factory-style assembly line, the e-waste industry must exhibit the following features[1]:

- Input: access to a regular, predictable supply of identical e-waste items.
- Process: use of repeatable standardized methods to produce output.
- Output: predictable output and revenues.
- Prices: fixed prices or predictable prices to buy input and sell output.

These are the ideal features that will enable e-waste processors to produce a high volume of output at lower unit costs. The average cost of producing output falls as the volume of output increases. In reality, the

---

[1] These features, at a conceptual level, are applicable to any industry and included in the economics and operations management textbooks. To adapt these for the e-waste context, I have referred to Goldhar and Jelinek (1983), "Economies of scale and scope" (2008).

e-waste industry does not exhibit these features. The way in which the industry operates *defies* economies of scale in e-waste processing.

The nature of input, process, output, and prices in the e-waste industry is structured such that it is impossible for an e-waste processor to realize economies of scale. E-waste supply is irregular and unpredictable, the products are not identical, and output and prices are not predictable. The way in which e-waste industry operates is based on *economies of scope*.

Scope is often equated with the idea of small, decentralized, specialist facilities with multi-skilled workers using a nonstandard, flexible, craft-based process on a wide variety of products.[2] These businesses are typically owned by close-knit communities, and the workers are often the owners. The process used is dependent upon the specific job in hand. The average cost of producing output does not fall as the volume of output increases. It is as economical to produce one unit as to produce many.[3] The features of e-waste industry, explained in the next few sections, illustrate *economies of scope*. Before we dive into the details, the assumptions made by the policymakers, stakeholders, industry leaders, and reality of the features of e-waste industry are summarized in Table 3.1.

Table 3.1 Assumptions and reality of e-waste processing industry

	Assumptions	*Reality*
Input	Regular, predictable	Irregular, unpredictable
	Identical, homogenous	High variety, heterogeneous
Process	Repeatable, standardized	Flexible, nonstandard
	Automation	Craft-based (manual methods)
Output	Predictable	Unpredictable
Prices	Fixed, predictable	Variable, unpredictable
Economics structure	Economies of scale	Economies of scope

[2] This idea of scope is synthesized from the traditional economic definitions (Goldhar & Jelinek, 1983; "Economies of scale and scope," 2008) and the concept of flexible specialization (Holmstrom, 1993).

[3] In other words, the costs of jointly producing a wide variety of products using the flexible process/methods are lower than when these products are produced independently.

INPUT: IRREGULAR, UNPREDICTABLE E-WASTE SUPPLY

Processors cannot ensure the steady supply of adequate quantities of e-waste to keep their facilities running without interruption. They also cannot plan *how much* and *when* they may be able to buy e-waste in the next month, quarter, or year. This is because e-waste supply is irregular and unpredictable. Consequently, processors cannot forecast revenues and plan resource requirements. As the owner of the informal processing company Aleph explained, he purchases varying e-waste quantities from different bulk consumers, yielding varying yearly revenues:

> During some years, we can earn Rs. 2 lakhs. During some other years, it becomes Rs. 3 lakhs. Some years, there are no earnings at all. Monthly, sometimes we earn Rs. 50,000. Sometimes, we earn Rs. 30,000. Sometimes, we also earn Rs. 1 lakh. We keep moving up and down. Sometimes, we earn. Sometimes, we don't earn. We do not have a straight level.

Informal processors such as Aleph manage this challenge by keeping costs down, operating a small roadside facility. The situation is difficult for large formal processors, who often operate at less than 25–30% of the installed capacity due to the lack of consistent e-waste supply. The business development manager of Epsilon, an informal-turned-formal processor, explained the difficulty associated with resource planning due to inconsistent input:

> Very often, what happens is, we will not have any work every 15 days because there is no material [e-waste]. We have to pay salaries on a monthly basis because labor [employees] will run away if we don't pay them salary. Sometimes, there is work. Sometimes, there is no work.

This resource planning challenge faced by processors due to inconsistent e-waste supply is compounded by the variety of e-waste types. E-waste supply consists of heterogeneous products, and this heterogeneity varies. A typical e-waste processor receives a mix of materials that ranges from aerospace to household electronics in a range of sizes: printers, fax machines, copiers, motherboards, printed circuit boards, CDs, floppy disks, tapes, cartridges, telephones, cell phones, telecom equipment, televisions, audio/video devices, dry cells, lithium batteries, microwave ovens, washing machines, medical electronics, scanners, magnetic resonance imaging machines, industrial, military/aerospace electronic equipment, enterprise servers, desktops, laptops, wires, tablets, etc.

This heterogeneity in e-waste impacts the processing costs. As explained by Epsilon's manager, recovering gold from gold-coated components and recovering silver from silver-coated components incur different costs, because the chemicals used are different. Disassembling a washing machine or refrigerator incurs a lower cost vis-a-vis metal recovery from circuit boards. Large-sized products such as servers incur lower disassembly costs when compared to small-sized complex products such as mobile phones. Epsilon's manager articulated the difference in effort needed to disassemble a server and a mobile phone:

> A phone needs lot of labor. It has many things that needs to be taken out and separated, the boards inside the phone have many small, small things, it takes time to separate those things. The bigger the electrical item is, the more easier it is to dismantle, smaller ones are difficult. For example, if I open this audio recorder, you will need to separate the board, speaker, they have metal and those metal parts are small, small things, it is labor intensive, actually. If we have a server, what is there in that? There are only cards in this big almirah type thing. Take out the cards, take out the plastic caps and send the boards for recovery. Hammer the almirah, and send it to scrap iron traders. So, material to material it varies.

Disassembling a server takes less effort and time vis-à-vis disassembling a mobile phone. This means processing costs vary depending on the type of product. When a processor cannot predict the *what type, when, how much* of e-waste they would get, resource planning is difficult.

PROCESS: FLEXIBLE, CRAFT-BASED METHODS

A repeatable standardized method cannot be applied to every e-waste product. The processor must assess every product based on the brand, potential for reuse and repair, and the value of commodities inside before determining the best course of action. There is a hierarchical order followed by processors in deciding how to process the e-waste. The first best option is reuse. If the product cannot be reused, the next best option is repair-cum-refurbish. If the product cannot be repaired or refurbished, it is cannibalized, and the parts are taken out. If the parts are not worth cannibalizing, the final option is to disassemble the product into commodities and recover metals through chemical treatment. A single processor does not do all these five processing operations; most will specialize in one or two of them.

While there is automatic machinery available intended to support e-waste processing, there are significant disadvantages to using this machinery. This machinery cannot be automated and applied to every product in a standard and repeatable manner. When the machinery is used, it reduces the quality and resale value of the materials it breaks down. Using automatic machinery to disassemble e-waste crushes the e-waste and separates it into different waste streams: ferrous (iron-containing) components and non-ferrous components such as copper, aluminum, and brass. These different waste streams, however, will have high levels of impurity compared to what can be achieved with manual labor. For example, copper gets into aluminum, and aluminum gets into copper. This impurity lowers the market price of both the copper and the aluminum. Commodity traders and recyclers pay a lower price for metals with impurities, as the contamination increases their processing costs.

Manual ways of disassembling e-waste generally provide maximum value, because manual repair and refurbishing can extract maximum value out of each product type. The manual method also enables processors to adapt to and derive revenue from the quantity of e-waste available at any given time. The expensive automatic machinery requires a high quantity of e-waste, typically more than 10,000 tons per year to operate economically. As we saw in the input section, obtaining a high quantity of e-waste supply consistently is elusive. On the contrary, the manual method is low cost and flexible to match with the varying low quantities of e-waste. The owner of RTH explained why the manual method is the most appropriate to operate efficiently in the e-waste processing industry:

> You have installed a machine [automation] worth Rs. 10 crores. You are not getting even 500 tons material per year. Total waste! You have made this much investment. For Rs. 10 crores, you will have to pay Rs. 10 lakh interest. If you cannot make even this, business model is a fail. If you ask about my business model, in the current scenario of market of e-waste industry, physical separation [manual method] is the best way to survive.

Investing in expensive automation makes economic sense only when there is a consistent high volume of e-waste supply. The high volumes would then offset the reduced revenues from impurities. As we saw in the earlier section, it is not possible to access consistent high volumes of e-waste supply.

Among informal processors, manual disassembly of circuit boards and other e-waste consists of craft-based methods that require deep knowledge of various products and parts. As we will see in Chapter 4, this technical knowledge is inextricably linked to the tight-knit networks and cultural protocols in which these processors operate. The value of this expertise is demonstrated by the demand for it among precious metal refiners. For example, Christian Hagelüken, who has more than 25 years of experience in precious metal recovery research and has held senior positions at precious metal refining companies, reported that metal recovery from circuit board components that are manually disassembled yields higher returns than the metals recovered from machine-operated components (Hagelüken, 2006). Precious metal refiners in the EU and elsewhere prefer to source circuit boards from emerging markets such as India and Africa ("Umicore, WorldLoop win award," 2014), because these markets have a significant presence of informal e-waste processors that manually disassemble circuit boards and parts from e-waste. This manual disassembly enables the precious metal refiners to maximize metal recovery.

Output: Unpredictable Due to Technological Change

The output of e-waste disassembly is unpredictable due to the technological change that is manifested in multiple ways: changes in the quantity of commodities, substitution of commodities, and miniaturization. Due to this, an e-waste processor cannot accurately forecast the revenues before processing the e-waste. For example, the owner of Aleph explained how the complex combination of heterogeneity in e-waste products and unpredictability in output negatively impacts revenues:

> The material [e-waste] that has come this time will not come next time. The material that will come next time, will not come the next time. Like this, products keep on changing, material also keeps on changing. There was some difference in the last PCBs [printed circuit boards]. Last time's PCB boards were different. The PCB boards coming now are different from the last time. Last time's PCBs used to fetch us Rs. 220. The PCBs

coming now, is fetching only Rs. 80, Rs. 70. There are no good metals in it.

In this case, the circuit boards recently received by Aleph fetched a lower price in the market due to lower quantity of metals. This wasn't the case with the circuit boards received earlier. In many cases, technological change has reduced the quantity of metals in electronics products. Technological change refers to the new—often improved and innovative—methods of producing and designing products (Jaffe et al., 2003). The quantity of commodities inside each product and, consequently, the yield that could be obtained after disassembling each product are unpredictable due to this technological change.

In some cases, metal content has increased or more expensive metal is being used. For example, the sales manager of a global circuit board manufacturing company explained that RoHS (the Restriction of Hazardous Substances) Directive adopted by the EU in 2003 has led to an increase in gold content vis-à-vis lead:

> The amount of precious metals in PCBs [printed circuit boards] have actually increased, especially due to RoHS. Earlier there used to be lead finishing. But, now due to RoHS you cannot have lead finishing, now it is gold finishing. So, the amount of gold has gone up by 60% because of RoHS.

The processors do not have complete knowledge regarding the potential output that can be disassembled from e-waste. This unpredictability is reflected across the industry when processors sell circuit boards to precious metal refiners and when processors buy e-waste from bulk consumers. For example, owner of AMY, a formal processor, overestimated the amount of metals inside and bought e-waste at a higher price:

> There have been cases where our valuation had gone wrong. We would have anticipated so much amount of metal in a material and would have quoted higher. After dismantling, we find that the quantity of metal is lower than we anticipated and we make losses. There is lot of unpredictability involved in this business.

Only after disassembling the e-waste, they realized that they have overestimated the amount of metals. This deal was a loss to AMY because the underlying unpredictability in output did not turn out in their favor.

New commodities with low recyclability are being substituted for recyclable commodities. If recyclability is low, processors cannot sell them to commodity traders and recyclers and cannot realize revenues. For example, Epsilon's manager explained the shift in the material used for laptop casing from plastics, which is recyclable, to Bakelite. Though Bakelite is lightweight, durable, and cheap and enables a strong casing relative to plastic material, it is non-recyclable and can only be incinerated (burnt at high temperatures). Unlike plastics, Epsilon cannot earn revenues by selling Bakelite. In another example, the managing director of an electronics component manufacturers' association quoted the case of metal cages used for telecom towers. Earlier, the cages were made of metals. But the metals have now been substituted with plastics. Plastics fetch a lower price when compared to the metals. This means the processors cannot earn revenues as high as was previously possible.

As electronic products get smaller, the labor required rises while the value of materials falls. Miniaturization, the trend of manufacturing smaller and smaller devices with better performance, has reduced the size of electronics products over time. According to Moore's Law, the number of components inside an electronic chip doubles every two years, thus reducing the product size and increasing the product's performance (IEEE, 2015). Smaller, lower-weight products mean lower quantities of commodities and, consequently, lower revenues for the processor. As the owner of RTH, a large formal processor, explained, the reduction in the size of products leads to a lower quantity of e-waste:

> The sizes have become compact every day. When you were buying P1, P2, P3 [computers run by Pentium processors] their sizes were big. Earlier, the laptop used to weigh 5 kg, now, the laptops weigh 1-1.5 kg. Earlier, the personal computer used to weigh 12 kg, now, it is 6 kg. Earlier, CRT monitors used to weigh 15-20 kg. CRT monitors have got over and now there is LCD monitor that weighs 1-1.5 kg. So, the size has reduced. Suppose, 5 years back you are collecting around 500 ton every month, if you were working with IT [bulk consumers who are global technology companies]. But today you will get only 250 ton.

On a per product basis, these technological changes have negatively affected the revenues of processors, whether they are informal or formal. For example, owner of Beet, an informal processor, said that this size reduction led to reduced metal content and lower profit margins:

> Earlier, even for doing scrap [disassembly into commodities] one used to get money. Now, there is not much money in doing scrap. Earlier, hard disks used to come in bigger sizes. So, metal content was higher. Now, size has become small. So, metal content has reduced and price is the same. But, after doing scrap you do not get that much profit as you were getting before. Earlier, the margin was like, you will get 50% to 70% as profit after dismantling. But now it has reduced to 10% to 5% profit.

Technological change manifested by changes in the quantity of metals, substitution of recyclables with non-recyclables, and miniaturization has led to unpredictability in the output and reduced revenues for processors. The challenges in input, process, and output of e-waste processing are compounded by the unpredictability in pricing.

PRICING: UNPREDICTABILITY DUE TO THE NATURE OF COMMODITY BUSINESS

Commodities are materials that are standardized, easy to exchange, have a fairly uniform price worldwide, and are the foundational materials used for producing many other products ("What makes something a commodity," 2017). Materials recovered from e-waste such as gold, copper, silver, plastic, and glass are classified as commodities. Regardless of whether these materials are recovered from a laptop, a washing machine, or a telecom tower, they have the same properties. Gold recovered from a laptop's circuit board by an informal processor is sold to a jewelry maker based on the international gold prices—spot price—set by the London Metal Exchange. The spot price of respective metals at the London Metal Exchange decides the price of respective metals disassembled from e-waste. For example, the owner of BB, an e-waste collection company, told a customer who came to enquire about disposing a non-functional desktop: "*If I promise an amount for the desktop today, and you bring the desktop next week, rates would have changed. Rates are based on the metal prices in the market.*"

E-waste processing is a commodity business. Processors view e-waste as a set of commodities bundled together. They unbundle these commodities by disassembling e-waste and sell these commodities to specialized commodity traders or recyclers. The price for these commodities varies daily, and processors face unpredictable prices when they buy and when they sell. If the product or its parts cannot be reused or repaired and sold in the second-hand market, the last option is disassembly into commodities. In this case, the revenues realized by processors are dependent on price movements in the international commodity markets. For example, when the informal processor Beet or the formal processor RTH disassembles their e-waste into various types of metals, the revenue they earn is dependent upon the prices of these metals in the London Metal Exchange. When Beet or RTH buys this e-waste from a bulk consumer, the price paid also depends upon these international market prices. In another example, an e-waste processor decides the buying price of a cathode-ray tube monitor, often found in televisions and desktop computers, based on the quantity of glass, plastic, and copper inside that monitor and at what price the glass, plastic, and copper could be sold to the commodity traders and recyclers. If the processor expects to sell glass, plastic, and copper from one monitor for ₹200[4] to the commodity traders on a specific day, then transportation, labor, and other processing costs are added, and the processor would be willing to pay ₹100 for one monitor on that specific day.

In summary, the nature of input, process, output, and pricing in the e-waste industry are structured such that the average cost of producing output does not fall as the volume of output increases. It is as economical to produce one unit as to produce many. Formal processors cannot operate profitably in such a context with expensive automation and large, centralized facilities. This is because the economics of e-waste industry is structured for economies of scope. What is then the best way to operate in an industry that is structured for economies of scope? To understand this, let's dive into operations management concepts.

---

[4] ₹200 is the value of the product based on the commodities inside and is popularly known as *scrap value*. This is irrespective of whether the product is functional or not. Processors calculate this scrap value for the worst-case scenario. If the products or parts in the e-waste purchased by the processor cannot be reused in the second-hand market, they can expect to earn a minimum by disassembling the products and selling the commodities inside. This logic of computing the price based on commodities inside is prevalent in transactions between stakeholders across the e-waste industry.

ECONOMIC CAPABILITIES TO EXCEL
IN INDUSTRIES WITH ECONOMIES OF SCOPE

Companies operating in any industry focus on developing specific economic capabilities[5] that help them operate and compete in their industry (Hayes & Wheelwright, 1984). The economic capabilities for a company are quality, time, cost, and flexibility (Hayes & Wheelwright, 1984). These capabilities are defined in various ways and can be high or low as appropriate to the features of the industry.

In the e-waste processing industry, these economic capabilities can be defined in the following ways:

- Low cost

  - Ability to operate at low costs by setting up small-sized facilities and leveraging manual processing methods. This reduces real estate and operational costs and enables businesses to weather lean times when there is no e-waste available.

- High flexibility

  - Ability to process a wide variety of e-waste items whenever they are available, at whatever quantity. This is achieved by employing multi-skilled workers who have the capacity to work on a wide variety of e-waste products.
  - Ability to employ on-demand labor depending on the e-waste availability.

- High quality

  - Ability to recover maximum value from e-waste through direct reuse, repair-cum-refurbishing, cannibalization, and disassembly into commodities. In the e-waste industry, this is possible by leveraging manual methods and high levels of technical skill and market knowledge.

---

[5] The term used in the operations management discipline is competitive priorities. To enable general readership, this is referred to as economic capabilities in this book.

- Quick turnaround time

  - Ability to disassemble e-waste and sell the commodities downstream as quickly as possible without storing them for a long time in a costly facility. This helps the processor to generate predictable revenues by selling the commodities at the expected price within the shortest possible time. The ideal situation is to purchase e-waste in the morning, process on the same day, and sell the disassembled commodities to downstream players by evening.

Informal processors' operating systems are developed in a way that naturally keeps the costs low, flexibility high, quality high, and turnaround time short. Informal processors operate in small, low-cost facilities, are multi-skilled, use flexible craft-based methods, and strive to complete e-waste processing in the shortest possible time. These approaches and conditions are aligned with the features of the e-waste industry that favor *economies of scope*.[6]

On the contrary, formal processors' operating systems are set up for *economies of scale* rather than *economies of scope*. When informal processors formalize, their operating systems' alignment with the e-waste industry is lost. An operating system that is naturally set up to realize *economies of scope* is forcefully shifted to realize *economies of scale*. This loss of alignment leads to *low cost* becoming *high cost* and *high flexibility* becoming *low flexibility*.

Formal processors' operating systems are typically characterized by high costs, low flexibility, low quality, and slow turnaround time:

- High costs: large-sized facilities (expensive real estate) with costly automation.
- Low flexibility: do not have multi-skilled labor to process a wide variety of e-waste; do not have easy access to on-demand labor.
- Low quality: do not recover maximum value from e-waste due to lack of technical skills, market knowledge, and reliance on automation.

---

[6] This is why, as we saw in Chapter 1, informal processors continue to be the dominant players in the e-waste processing industry and their operations are naturally structured to create value from e-waste.

- Slow turnaround time: store the disassembled commodities for a period in expensive real estate and wait to sell when the commodity prices are favorable.

Formal processors with these operating systems in place regularly experience low profit margins, lack of enough e-waste supply, low capacity utilization, and other negative impacts on their businesses. These approaches, which are predicated upon *economies of scale*, are not appropriate for the context of India's e-waste industry.

Throughout my fieldwork, many formal processors and informal-turned-formal processors shared experiences signaling their operating systems' lack of alignment with the e-waste industry. The manager of CIT, a formal processor, said: "*E-waste supply is not steady. We cannot afford to staff labor daily. Hence, we accumulate the material to a certain level and call labor for dismantling.*" Here, we can observe the low flexibility stemming from lack of access to on-demand labor to manage the irregular e-waste supply. CIT hires informal processors to do the e-waste processing in their formal facility. The owner of Delta, an informal-turned-formal processor, who spoke about the constant pressure to acquire more e-waste after becoming formal:

> Now, we have become formal. See, if we go home and sleep, we keep thinking about the monthly rent, labor charges. These kinds of tension were not there before. Bring material, hire 4 laborers for 4 days, do the work, finish the work, then go and bring another material, it was like this [in the informal]. Now, only if you give work continuously, laborers will stay with us. If you give work for 4 days and do not give work for another 4 days, how will laborers stay with us? I must sit and work myself.

The owner of Delta did not face this pressure during his days as an informal processor because his operating system was aligned with the irregular unpredictable e-waste supply. His costs were low, and he could hire on-demand labor to process e-waste. After setting up Delta, he could no longer afford to operate in the same way and, instead, had to ensure consistent e-waste supply to keep up with the higher operational costs. Delta's transition to formalization broke down previous alignment with their market context.

As we saw in Chapter 2, many formal processors adopt "business-as-usual" practices to maneuver within India's e-waste industry. The misalignment between formal processors' operating systems and the realities of India's e-waste industry creates a need for these workarounds. The realities of India's e-waste industry do not support the operational systems of formal processors, and these businesses subvert ostensible best practices and regulations to improve flexibility, lower costs, and speed up the turnaround time. When formal processors sell a portion of e-waste to informal processors due to a lack of capacity or skilled labor, this signals the effort to improve flexibility. When a formal processor sells the disassembled commodities to an informal commodity trader who provides a prompt service and pays money immediately vis-à-vis a formal commodity trader, this signals the effort to speed up the turnaround time.

Within the current market conditions and policy context of India's e-waste processing industry, formal processors rely on informal processors to meet their economic and operational needs. To thrive in the e-waste industry, the informal processors' operating system should be adopted by the formal processors. Some formal processors have already absorbed this lesson, and they have begun to develop capabilities like those of informal processors by, for example, investing in small-sized, low-cost facility and operations without investing in expensive assets such as automatic machinery. To date, there has not been enough research or communication for formal processors to understand the practices that will serve their businesses best within the current system.

Conclusion

India's informal processors have managed to thrive within the constraints of policies that intentionally limit their operations. This is because their inherent economic capabilities are aligned with the *economies of scope* of e-waste industry. Informal processors are highly flexible and low cost, recover maximum value, and can offer speedier turnaround times than their formal counterparts. Because of these factors, they can purchase e-waste from large companies at higher prices, incentivizing those companies to dispose of their e-waste, and create new value for buyers seeking affordable, refurbished electronics. These working mechanisms, in turn, enable access to affordable computers across India's school system, help low-income households acquire more affordable electronics, and extend the useful life of electronics products. For example, the owner of the informal processer Aleph explained that routing functional computers into the second-hand market makes it affordable for low-income households to further children's education:

> If 100% material is worth reusing, it is reused. If it is reused, who will get benefited? See, I have money, you also have money. You can afford to buy CPU, monitor for your kids by paying ₹40,000 - 50,000, your kids will use it and learn. Now, there are these small, small people with a salary of ₹10,000 - 15,000, those people need to spend that money for their living expenses. If those people get material [computers] through second-sale, their kids are benefited, doing this will improve our country.

In addition to making computers affordable, the economic capabilities of informal processors also help to extend the useful life of other electronics products. Cool, another informal processor based in Bengaluru, collects customer-returned products from electronics retailers. Cool does repair-cum-refurbish, cannibalization operations and second-hand resale for products such as washing machines, mixer grinders, refrigerators, televisions, and other household electronics. They sell these second-hand products to both direct consumers and second-hand dealers with wider geographic distribution channels.

The informal processors' inherent economic capabilities that extend the life of electronics products are built on the foundations of social networks and deep expertise.[7] Relationships with the informal processors enable schools and service centers to acquire affordable products and spare parts. The owner of Beet, an informal processor, spoke about the relationships with schools and service centers:

> There are some people from schools, some from service centers. They let us know their requirements, we inform them when we get material. They will come, see if there is something functional that can be used as spares, if there is something that can be reused they will buy these things.

As we will see in Chapter 4, the unique cultural context underlying India's informal processors is a dynamic and often ignored strength within the broader e-waste industry. By adapting and designing policies that leverage

---

[7] As we saw earlier in this chapter, *economies of scope* have economic and cultural aspects. This chapter explained the economic capabilities and the next chapter will explain the cultural aspect such as the role of close-knit community networks and deep expertise that translates to superior craft-based skills in e-waste processing.

the strengths of that existing system, decision makers can foster the conditions for both formal and informal processors to adopt safer and more sustainable practices without sacrificing cultural heritage or economic benefits.

REFERENCES

Economies of scale and scope. (2008, October 20). *The Economist*. https://www.economist.com/news/2008/10/20/economies-of-scale-and-scope

Goldhar, J. D., & Jelinek, M. (1983). Plan for economies of scope. *Harvard Business Review, 61*(6), 141–148.

Hagelüken, C. (2006). Recycling of electronic scrap at Umicore precious metals refining. *Acta Metallurgica Slovaca, 12*, 111–120.

Hayes, R. H., & Wheelwright, S. C. (1984). *Restoring our competitive edge: Competing through manufacturing*. Wiley.

Holmstrom, M. (1993). Flexible specialisation in India? *Economic and Political Weekly, 28*(35), 82–86.

IEEE Spectrum. (2015). *The law that's not a law*. IEEE. https://ieeexplore.ieee.org/document/7065416

Jaffe, A. B., Newell, R. G., & Stavins, R. N. (2003). Technological change and the environment. In K. Mäler & J. R. Vincent (Eds.), *Handbook of environmental economics: Environmental degradation and institutional responses* (pp. 461–516). Elsevier.

Stahel, W. R. (2016). Circular economy. *Nature, 531*, 435–438.

Umicore, WorldLoop win award: Belgian metals producer is recognized for working to boost electronics recycling in Africa. (2014, October 29). *Recycling Today*.

What makes something a commodity. (2017, January 3). *The Economist*. https://www.economist.com/the-economist-explains/2017/01/03/what-makes-something-a-commodity

Specialized Social Networks with Deep Expertise: The Value of Cultural Capital in India's E-waste Processing Industry

Abstract To develop effective policies to govern e-waste management in India, it is critical that we understand the cultural factors that shape this industry. Informal processors have developed effective systems over a period of centuries that yield tangible economic and cultural gains. These benefits are driven by a tight-knit network of community-based professionals who share not only intergenerational technical expertise, but also sophisticated networks and shared values that facilitate a reciprocity-based economic system. This chapter will surface and identify the specific traditions and processes that make this unique system effective. As we dive into this cultural context, we begin to see that informal processors are indeed the mainstream and not a residual or a fringe alternative to the formal economy.

Keywords Informal recycling · E-waste recycling · History · Cultural capital · Green skills

INTRODUCTION

In October 2014, I traveled to the outskirts of Bengaluru for a conversation with the owner of Zeta. Zeta is an informal-turned-formal processor that developed as a spinoff from Beta, another informal-turned-formal processor co-founded by the same owner. I visited during a lean period, and the Zeta facility was empty of both staff and e-waste materials. My interview in the owner's office was punctuated by back and forth calls with another informal-turned-formal processor as the two owners coordinated bidding for an online e-waste auction. While the auction was designed to direct e-waste to the highest bidder, these two enterprises worked in collaboration to build a consensus and make sure that one of them could obtain the materials. This way, they could keep the e-waste and future profits generally within the fold of their community. This is one of many ways that e-waste processors in India tap into their networks, share knowledge, and coordinate mutual support.

Coordinated efforts enable small enterprises in India's e-waste sector to amplify limited resources with community-wide returns. When Alpha's owner was unable to fund the e-waste purchase from a large company, for example, he reached out to informal processors and informal-turned-formal processors in his neighborhood community (Reddy, 2011). They contributed modest amounts of money and collectively purchased the e-waste. The total profit from processing this e-waste was distributed to each member proportionate to their contribution. These relationship-driven business practices are common among informal processors and informal-turned-formal processors who have deep roots in the informal economy. While many formal processors are baffled by the inner workings of these negotiations, the practices themselves are critical to the functioning of India's e-waste industry.

We saw in Chapter 3 that the economic capabilities of India's informal processors are well suited to support an e-waste processing industry built for *economies of scope* rather than *economies of scale*. To illuminate how these business models work in context, this chapter explores the many cultural factors that shape this industry and shows how these factors reinforce capabilities and support collective thriving within e-waste processing communities. Generations of cultural and industrial evolution have shaped informal processors' highly-specialized social networks, driving relationship-driven business practices. This cultural capital[1] yields deep expertise that makes the informal processing system effective. To develop effective policies to govern e-waste management in India, it is critical that we understand the cultural context of India's e-waste processors and identify how these existing systems offer pathways to economically viable and environmentally sustainable policies.

HISTORICAL CONTEXT
OF INFORMAL E-WASTE PROCESSORS

To create a better future for India's e-waste industry, we first need to understand the connections between its past and present. Many of India's informal e-waste processing roots can be traced back to the sixteenth-eighteenth centuries, when the artisans made metal crafts during the Mughal Empire that controlled a major part of South Asia (Roy, 2009). For instance, the brassware manufacturing industry in Uttar Pradesh's Moradabad city has existed there since the seventeenth century (Roy, 2009). Moradabad is also a core hub for informal processors' metal recovery from e-waste (Doron & Jeffrey, 2018), and it is likely that these metallurgy skills share their roots with the brassware industry. There is archaeological evidence for the exceptional workmanship and processing of various metals such as copper, gold, silver, and iron, as early as the Harappan Phase of the Indus Valley Tradition, from 2600 to 1900 B.C. (Kenoyer & Miller, 1999). During this period, metals were mined from ores, melted, and alloyed with other metals for improved properties. Archaeological evidence also shows that scrap metals were collected, melted, and recycled to make new products (Chakrabarti & Lahiri, 1996). Other historical examples such as the rust resistant Iron Pillar at Delhi, the Iron Beams of Konark Sun Temple at Odisha, and high-grade Wootz Steel at South India are outstanding achievements recognized by metallurgy scholars around the world (Srinivasan & Ranganathan, 2013).[2]

---

[1] The term "cultural capital" is borrowed from Bourdieu (1973) and translated to the e-waste context. It is the idea that the informal processors are embodied with the assets such as specialized social networks and expertise that lend the capabilities to succeed in this industry.

[2] A study in the National Metallurgical Laboratory's Technical Journal linked the rust-resistant composition of the Delhi Iron Pillar to the present methods practiced by the Indigenous community (known as Adivasi community) who live in remote forest areas

Indeed, the world's most prominent metal refiner, Umicore, has linked its leadership in metal recovery from e-waste and other industrial products to its roots in mining and metallurgy (Murray, 2013). These histories and others like them have contributed to the rich culture that underlies e-waste processing in India today.

Traditions like these have facilitated a highly-developed network of experts who share and transmit advanced technical knowledge about the processing of modern electronics through culturally-specific mechanisms of knowledge transfer. This long history of metalwork has evolved with the technologies of the materials themselves. For example, one of the core social networks of today's informal e-waste processors in Bengaluru traces its roots to the Kolar Gold Fields, where their ancestors were recovering gold from mine runoff in the nineteenth century. The Kolar Gold Fields was formally established in the colonial era in 1880s, though gold mining was being carried out much more than 100 years before the formal establishment (Britannica, 2015; Srikumar, 2014). Through trial, error, and ingenuity, the ancestors of today's informal processors developed acid leaching processes to extract gold from the mine runoffs. As the gold in the mines began to dwindle, these ancestors began migrating to other parts of the state and seeking out alternative pathways for their technique. The leaching technique used in the mine runoffs became a popular method to extract gold from gold-coated watches and pens.

In the case of the Kolar Gold Fields, these artisanal miners developed technical expertise in gold recovery and honed their craft. The techniques evolved alongside technologies themselves and were adapted over time first for telecom equipment and, later, to computers and circuit boards. Over more than three generations, this process has translated into a wide network of ventures across the informal e-waste ecosystem in Bengaluru and made it a thriving hub for informal e-waste processing. Specific applications and processes for the original leaching methods have diversified across communities and geographic locations with social networks specializing in extraction from discrete types of e-waste, such as transistors, and materials such as brass and copper. Today, all the major cities in India

(Lahiri et al., 1963). In this paper, the authors had asked the Adivasi community to make iron using their Indigenous methods and had compared this with the microstructure of the existing Delhi Iron Pillar. The authors reported the striking similarity in the microstructural properties of the iron that has helped Iron Pillar resist rust for 16 centuries.

have their own scrap recycling hubs. This progression of innovation has developed from testing and refining of technical processes based on new discoveries, new materials, and the technological changes.

In these specialized social networks, kinship ties are essential to business collaborations and the transmission of expertise across generations. Apprenticeships are only available to those with family ties or, on occasion, strong ties through a shared religious community. During my fieldwork, informal processors, informal-turned-formal processors, large formal processors, and commodity traders and recyclers discussed religion in the context of their work. The topic of religion and its business relevance is widely acknowledged in India's e-waste processing industry. Many stakeholders understand and acknowledge the prominence of Muslim community in informal e-waste processing.[3] All of the informal-turned-formal processors and the informal processors I met during the fieldwork identified as Muslims through their names, attire, and residential neighborhoods. The owners of Alpha and Mash (both informal-turned-formal processors) received their training in e-waste processing more than three decades ago through apprenticeships with experienced informal e-waste processors who belonged to the same religious community. The owner of Amy, a formal processor in Maharashtra, has kinship ties with the Bengaluru-based Alpha. Amy's roots are in the informal processing, and they continue to do business with Alpha.

The closeness and exclusivity of these religious ties has historically been a determining factor for professional pathways in India (Harriss-White, 2003). Harriss-White (2003, p. 148) describes the dominance of Muslims in craft-based professions ranging from brassware to hand-printed textiles and perfumes to silkworm rearing as follows:

> In Uttar Pradesh, for example, Muslim artisans produce brassware in Moradabad; pottery in Khurja; glassware in Ferozabad; carpets in Bhadodi and Mirzapur; carpentry and woodwork in Sharanapur; hand printed textiles in Farrakhabad; cotton and silk embroidery in Varanasi; perfume manufacturing (to which the development of *unani* medicines is related) in Lucknow, Kanauj and Jaunpur; and handloom cloth in Mau. In Bihar, large numbers of Muslims are silk and cotton handloom weavers. Muslim

---

[3] Though e-waste processors in the informal economy predominantly belong to Muslim community, anecdotal evidence indicates there are informal e-waste processors from other religious communities too.

workers dominate the *bidri* ware and carpet industries in Andhra Pradesh, and silkworm rearing and toy industries in Karnataka.

The dominance of Muslims in varied craft-based professions exists across India. Professions associated with waste recycling—animal bones, paper, metals, plastic, rubber, etc.—have traditionally been taken up by Muslims (Harriss-White, 2003). This tradition has likely played a role in the prominence of Muslims in the e-waste processing industry today. The business practices of informal e-waste processors follow the same protocols and approaches as many other professions in this context. One Muslim business owner, for example, described the application of his previous profession as a butcher to his current e-waste processing work as the owner of Mash (Reddy, 2011):

> Do you know what happens in the lamb business? People in the lamb business purchase the lamb on a wholesale basis but sell it on a retail basis: i.e., lamb butchers say bismillah and kill the animals and soon after they set out to divide the lamb into various parts. Very soon there are separate piles for skin, intestine, feet and legs, head, and meat for household consumption. These piles are then auctioned off separately to vendors that deal in those individual parts. This is how they make a profit. Having seen the lamb business is how I learned to do computer business as well.

Just as there are markets for the individual parts of lamb in the meat industry, there are markets for individual parts of computers in the e-waste industry. The owner of Mash applied his knowledge of how markets work among butchers' distribution of lamb parts to the afterlives of technology.

The culturally-driven specialization of professional niches has facilitated a highly organized e-waste processing industry. This industry has demonstrated resilience in the face of external changes in market forces, new technologies, and global policies that pose barriers to business growth. At the same time, the cultural divide between informal e-waste processors and other stakeholders creates unique industry challenges. During my fieldwork, bulk consumers, large formal processors, and commodity recyclers noted the difficulty of doing business with Muslim informal processors because of their business approaches and reluctance to admit outsiders. As one commodity recycler (who also does informal e-waste processing occasionally) noted, "*I find it very difficult to do business with these Muslim recyclers.*" For informal processors who are Muslims, this cultural divide can create additional obstacles. Many followers of Islam, for instance, do not accept loans or interest rates on spiritual and cultural grounds. The manager of Epsilon, for example, is a Muslim and has completed a post-graduate management degree in a prominent Indian business school. When I asked about whether he had approached banks for business loans, he said: "*I don't go for loans; I don't believe in interest. I neither take interest or pay interest. When I bought a Harley [Harley-Davidson motorcycle], I paid cash. If I have cash, I do my business. Because, interest is forbidden in our religion.*" Current formalization policies do not offer culturally-appropriate alternatives to invest in equipment and other major expenses that could move businesses forward. This hurts the ability of business owners from Muslim communities to operate at low cost, which limits their ability to do business in compliance with formalization policy.

India is a religiously pluralistic country with more than 1 billion people and a long history of complicated social politics, including the contested caste system. While there is often resistance in the Global North to linking conversations about religion or social status to corporate practices, these dimensions are critical to understanding and doing business with India's e-waste sector. Processors, bulk consumers, commodity traders, and recyclers in the e-waste industry understand the importance of social communities, which are often tied together by a common religion or caste. There are such social communities (such as Marwari Hindus) involved in industries such as plastic recycling and scrap metal recycling, and these community-based industries have become more prominent over time. For example, the founder of a scrap metal importing company MMT is a Hindu and began importing scrap metal during mid-1970s. Over the years, members from the Marwari Hindu community entered this business because the perception of recycling industry changed from *waste* to *resources*. In his words:

In old days, this kabaadi [scrap] business was dominated by Muslim people, Marwaris were not there, it was mostly done by Muslims. Marwaris came later. All these trash here and there, were picked and traded by Muslims. Now, what has happened, in all places educated people are also coming in this business. Why? Now, recycling industry is a means for raw material of the industries, the concept has changed, you understand? Recycling industry is doing job for the raw materials, they are importing not scrap, but raw materials, so, this scenario changed.

Muslims have historically dominated professional niches such as waste recycling that are not traditionally taken up by other communities (Harriss-White, 2003). Gradually, when the perception of recycling shifted from *waste* to *resources* and *raw materials*, members from other communities entered the industry.

The point to be understood from this discussion of religion, caste, and business is that there is a network of traditions and relationships beyond the faith itself at play in the e-waste processing industry. Recently, scholars have studied the significance of relationship-based business practices in community networks in the working of other industries in India (Vaidyanathan, 2019). For example, Tiruppur is a world leader in knitted garments manufacturing, an industry dominated by the Gounder community; Surat cuts and polishes more than 90% of the world's diamonds, and this industry is dominated by the Patel community (Vaidyanathan, 2019). India's e-waste processing sector is one of many recycling industries where these community dimensions are critical to the development of effective policies and business practices.

Tacit Knowledge and Embeddedness

These specialized social networks with deep expertise have developed through two core mechanisms: *tacit knowledge* and *embeddedness*. *Tacit knowledge* is knowledge which cannot be codified or made explicit in written or verbal form (Lam, 2000; Polanyi, 2009). It cannot be shared because it is built into the context. Polanyi (2009) describes human being's ability to recognize faces as tacit knowledge: "*We know a person's face, and can recognize it among a thousand, indeed a million. Yet we usually cannot tell how we recognize a face we know. So most of this cannot be put into words.*" In India's e-waste sector and broader recycling industries, tacit knowledge enables informal processors to answer questions such as: At what price should the e-waste be purchased? What materials can be extracted from a given electronic device? What is their worth in the current market? Who would buy them? Which processing operation will yield the highest return on investment? These are the million-dollar questions in the e-waste processing industry, and informal e-waste processors can answer them with reasonable precision by visual inspection alone.

While tacit knowledge is often informed by industry experience, it is distinct from other forms of technical knowledge by its difficulty to be expressed explicitly. I encountered one of the most striking examples of tacit knowledge in my fieldwork when I visited a small-scale plastic recycler on the outskirts of Bengaluru. This recycler converts plastic waste to plastic granules and sells them to large plastic product manufacturers. The two employees of this small-scale company tested the quality and grade of plastic granules by biting into them with their teeth, rather than using any equipment. I was given two samples of plastic granules neatly covered in a transparent plastic cover. Visually, I could not distinguish between the two samples, and both seemed alike. One sample was low density plastic (LD) and another sample was polypropylene (PP). LD and PP have virtually identical visual characteristics and exterior textures. Employees at this facility distinguished between the two materials by biting into 2–3 granules from different one-ton sacs. Using their teeth, they were able to discern differences in hardness and other characteristics that are difficult for outsiders to know. It is through these manual tests that business deals are done.

Formal processors will pay a premium to a get reasonably precise answers to their questions regarding e-waste assessment and pricing. That's why, as we saw in Chapter 2, many formal processors rely on the technical and business knowledge of informal processors to move their businesses forward. When I met jointly with the business development executive of a large formal processor and owner of Aleph (an informal processor), both spoke about their shared business pursuits. The owner of Aleph accompanies the business development executive to visit warehouses of bulk consumers for e-waste assessment and pricing. The formal processor gives a commission to Aleph for providing this service. Unlike most formal processors, Aleph's knowledge of e-waste assessment and pricing is nearly instinctive. While I was at the facility of Cool (informal processor), its owner calculated the price at which he would buy my laptop. He asked me for a few details such as the brand name, configuration, age, battery back-up, in-built web camera, whether the edges are damaged, and whether the laptop had any scratches. Based on these details, he said he could sell this laptop at a specific price in the second-hand market and added his margin to calculate the price he would pay me. In less than a minute, the owner of Cool was able to assess the value of my laptop based on his knowledge of the product, the downstream market, and buyers.

Apart from possessing this technical knowledge on e-waste assessment and pricing, informal processors excel in disassembly and metal recovery. For example, CIT—a large formal processor based in Bengaluru—hires informal processors and invites them to CIT's formal facility to break down the e-waste using their deep expertise in disassembly. Consultant X recounted a business visit to Europe during the mid-2010s with representatives from the informal economy. Whereas the European formal processors discussed the technical challenges to recover platinum from a specific e-waste material, the informal processors already had the expertise to identify the solution. They said: " *We have been doing it for the last ten years. Will you export it to us? We will do the job work for you and send it back.*" This knowledge is intentionally guarded, resides within the specific families, and is not shared beyond family members.[4] For example, the owners of Beta, Delta, and Zeta belonged to this family, all of them specializing in metal recovery. Apart from possessing this technical knowledge on e-waste disassembly and metal recovery, informal processors also excel at integrating new technologies into existing systems. For example, the informal processors who chose to formalize made the dust collection machinery and other safety equipment by themselves without relying on an external vendor, thereby saving costs and helping other members to develop this technology.

This tacit knowledge in e-waste processing intertwines with rich networks of *embeddedness* that enable informal processors to form specialized social networks. *Embeddedness* refers to the business practices related to community norms and relationships, rather than market forces only (Granovetter, 1985). This was first introduced as a scholarly concept by Karl Polanyi (1944), who used the term to define the inextricable relationship between economic activities and kinship, religious, and political institutions in pre-industrial non-market economies. Granovetter (1985) later extended this term to define such inextricable relationships in market economies. Granovetter (1985) explained that, in certain markets, the business activities are embedded in concrete social relations and networks of such relations who closely monitor each other's behavior to improve trust and reduce malfeasance. The role of embeddedness is central to the success of the informal e-waste processing industry across India. In the context of India's e-waste industry, embeddedness and tacit knowledge are woven through a tight network of informal processors who share religions, dialects, and cultural protocols. This has enabled them

---

[4] Regardless of the available expertise, existing policies and stigma block the informal processors from creating value. As we saw in Chapter 1, informal processors cannot obtain the government approval to import e-waste or export the recovered materials.

to coordinate business activities based on trust-based social relationships. As we saw in the opening paragraphs of this chapter, embeddedness plays a critical role in business activities such as purchasing e-waste. The informal processors and informal-turned-formal processors sell the processed items to downstream second-hand markets and commodity traders based on these social relationships. The information about these downstream buyers is kept as trade secrets, not documented, and not revealed to outsiders.

The workers who bite plastic granules to determine their makeup, the owner of Cool who can assess a laptop's value in less than a minute, and the informal processors who offer their platinum recovery capability to the European players share a key set of challenges. Like so many other informal processors, these operations lack access to institutional finance, legitimacy, and favorable media coverage that are available to their formal counterparts. They succeed, instead, through their deep expertise and specialized social networks of business collaborators who collectively leverage their niche areas of expertise. Across India's e-waste industry, various stakeholders recognize informal processors for their social networks and expertise for e-waste processing. During industry conferences, policy makers and formal processors speak highly about the skills of the informal processors. But this recognition has not translated to inviting informal processors to these conferences, actively listening to their views, giving them a seat in the decision-making table, or engaging them to co-create e-waste laws.

India's informal processors thrive because they have cultivated rich social networks and cultural protocols that facilitate not only exceptional technical expertise, but also positive social and economic impacts for entire communities. As we will see in Chapter 5, recognizing the strength of informal processors is a critical first step to unlocking their potential in policy futures. We can better appreciate these policy futures by understanding the ways in which informal processors are able to succeed despite the barriers that current policies have created.

BUSINESS PRACTICES THAT ENABLE COLLECTIVE RESILIENCE

The success of the informal processors in the face of barriers can be attributed to their abundance mindset and minimalist lifestyle. Their business practices reflect a commitment to supporting the community as a whole. They have an *abundance mindset* rather than a *scarcity mindset* (Covey, 1989). They do not view e-waste business as a finite pie where one processor taking a large slice will leave less for everyone else. Rather, they view e-waste business as an infinite pie with plenty of slices for every processor. For example, I visited Beta's facility only to find that it was temporarily closed due to lack of e-waste supply. Facility closure meant zero revenues. This lean period notwithstanding, the manager of Beta demonstrated an abundance mindset:

> If you go to market, there are so many vegetables vendors, so many shops are there. God will give business to everyone. Like that, we are thinking. Now, presently 50 recyclers are there, if 100 recyclers come, God will give the business. If our destiny is good, definitely we will be good.

What motivates them to work this way, and why is this model effective? Conventional competitive models don't work in this context or align with the values or approaches of many informal processors.

As we saw in the opening paragraph of this chapter, informal processors shared the wealth by collaborating to buy more e-waste, rather than seeking to gain as individuals. This dynamic is also demonstrated when informal-turned-formal processors are aware of the operations and whereabouts of the informal processors. In a competitive way of running a business, these informal-turned-formal processors might increase their market share by closing down the informal processors' operations by revealing them to the government. But they do not do that due to their value system. The owner of Gamma, for example, said: "*If we go and complain, it will make their life difficult. These are small people earning some 100-500 rupees. Even God will not forgive us, if we do that.*" This value system is also good to move their businesses forward.

In another instance of this abundance mindset, the owners of Gamma and Delta proposed that bulk consumers should not get into long-term multi-year agreements with a single formal processor. The current practice in e-waste business is that a large bulk consumer would sign a long-term 3-year or 5-year deal with a formal processor. This means that for the next 3–5 years, all the e-waste is sold only to that formal processor. This practice prevents other processors from purchasing e-waste and moving their businesses forward. In this context, it would be more appropriate for bulk consumers to enter yearly contracts with one processor. For example, during year 1, the e-waste is sold to one formal processor, during the second year, e-waste is sold to another processor, etc. This way, all formal e-waste processors (formal and informal-turned-formal) will benefit. The owner of Gamma, for example, said: *"Don't give only to one person. I'm telling you, don't give only to me. Give to all 43 recyclers. There will be 100 recyclers in the coming years. Give to all 100."* The informal economy wants all processors (informal or formal) to prosper and do well because they care about each other's livelihoods and recognize that it is collectively good for business. They believe in cooperating, rather than competing, with other informal processors, informal-turned-formal processors, and even the large formal processors. This enables informal processors to leverage their economic capabilities such as low cost and high flexibility, maximizing economies of scope.

This abundance mindset extends to creating social impact for the entire community. During my research, I was at a community center office in Bengaluru's e-waste processing neighborhood. This community center was run by Beta's owner, and the profits from e-waste business were redistributed for education and life events for the poor.

Beyond this abundance mindset, informal processors have a minimalist lifestyle focused on quality of life. The mentality of informal processors is to work for few days a month and then relax for rest of the month. Even if they could earn a profit of minimum ₹5,000 per day, many do not want to work during the entire month only to earn more money. Rather, they work for 5 days a month and earn enough money to meet their priority needs, and the rest of the month is spent with their family and playing with their kids. As Consultant X described, the informal processors' mindset is different from the conventions of the formal corporate world:

> Their mentality was: 'why should we work all thirty days in a month?' [...] And once they get that [minimum profit], then rest of the month, they won't work, they will sit at home [...] The rest of the days they are playing with their kids and they are watching TV and when the money is over, then again they start working. But whenever they want they used to get the material, how much ever they want.

Consultant X once tried to motivate the informal processors to work for more days a month so that they could earn more money. They replied by asking what they would do with more money, as they were not interested in buying a big house. They were content with their small houses, and they prioritized having a good TV, mobile phone, watches, etc., and were able to meet their priority needs by working 5 days a month. Specifically, they responded to Consultant X by stressing the importance of leisure and family to leading a meaningful life:

> How many days do you work? You studied all these years and then you earn, and then everyday you work, and then when do you live? [...] When do you get time to sit and watch TV? When do you get to sit with your children to play?

Their responses to Consultant X left her reflecting deeply upon her own life, education, and career. She concluded that, when compared to herself, the informal processors seem to lead a happier life:

> The priorities, they have one set priority and they are happy. Finally, what are you living for? You need to be happy. After studying all this, after working, slogging like this, if you ask me, if you compare me with one informal sector, they are much more happy. [...] So, this is one social thing which I have learnt from them, I don't have an answer, it's something very very amazing.

Many informal processors who transitioned to formal spoke about higher levels of stress and loss of happiness in life that came with their new formal status. Informal processors who had been contentedly leading happy lives by working 5 days a month now had to constantly worry about buying more e-waste to run their facilities at full capacity every day. Consultant X has noted this difference in the informal-turned-formal processor Beta: "*This difference is due to the formalization of his business. In fact, his happiness index would have come down due to formalization.*" The informal-turned-formal processor Delta, for example, launched a successful formal enterprise by leveraging preexisting relationships. Economic gains notwithstanding, the owner of Delta cited higher levels of stress after formalizing: "*We were happy in the informal too. Now, we have become formal, if we go home and sleep, we keep thinking about the monthly rent, labor charges. These kinds of tension were not there before.*" Both Aleph and Beet made the transition from informal to formal, but both returned to their informal businesses after becoming formal and are happy with their current lives. Unlike Delta, they did not need to worry about frequently purchasing their next e-waste supply or paying high rentals for the facility. After resuming their informal businesses, Aleph and Beet were operating with low cost, high flexibility, high quality, and quick turnaround time capabilities that are aligned with the nature of the e-waste industry.

The emphasis on formalization has put many of the informal processors in a position of having to choose between way of life and livelihood. As we have seen, formalization policies biased toward scalability have created barriers and pressures within India's e-waste industry. If we reorient our approach to facilitate economies of scope, there is high potential to co-design models that honor cultural protocols, amplify existing capabilities, and enable sustainable and profitable solutions for informal processors. Such models will provide greater flexibility for informal processors to contribute to the e-waste industry without sacrificing their traditions, lifestyles, and community-driven approaches.

CULTURAL CAPITAL ENABLES THE ECONOMIC CAPABILITIES OF INFORMAL PROCESSORS

As we saw earlier, India's e-waste processing industry cannot function without the deep expertise of its informal processors. The rich historical context that enables embeddedness and tacit knowledge shapes business practices in ways that diverge from mainstream operating practices in the formal sector. The economic capabilities of informal processors rest on the foundation of cultural capital. For example, the specialized social networks of informal processors translate to the capabilities of *quick turnaround time* and *high flexibility*. An informal processor who leverages embeddedness can quickly find appropriate downstream buyers in the second-hand market or commodity trading market. This translates to speedier turnaround time without storing significant inventory or waiting for the right buyers. An informal processor who leverages embeddedness can also hire skilled labor on-demand. The deep expertise of informal processors translates to the capabilities of *high flexibility* and *high quality*. An informal processor who leverages tacit knowledge in processing multiple e-waste items can process a wide variety of e-waste (high flexibility) and recover maximum value by choosing the appropriate processing operation (high quality). Cultural capital enables the informal processors to efficiently process a high variety of products in the e-waste stream and convert them into forms usable for the society.

The wider informal economy in the e-waste supply chain consists of informal processors, informal retailers in the second-hand electronics market, and informal commodity traders and recyclers. These enterprises make up a network of independent small-scale companies often run by close-knit culturally homogenous communities. They consist of multi-skilled workers who often use manual methods, have business practices based on social relationships, and can quickly reconfigure operations according to rapidly changing market forces and demand. Rather than a group of large companies competing by mass-producing a standardized product in high quantities using assembly line workers, here we have a group of small community-based enterprises (specialized social networks) reusing and repurposing a wide variety of e-waste items using deep exper-tise (tacit knowledge). These small community-based enterprises cope with complex set of market forces where the final outputs from e-waste processing and their prices are variable and unpredictable.

This system has enabled the informal processors to develop capabili-ties to process a wide variety of e-waste items in an efficient manner by maximizing the economies of scope we saw in Chapter 3. The framing of current scholarship and policies, drawing from the models of devel-oped countries, often suggests that informal economy is a residual or a fringe alternative to the formal economy (La Porta & Shleifer, 2014). A closer look at the context of India's e-waste industry, however, shows that informal processors represent the mainstream.

While some business and management scholars have begun to recog-nize the strengths of informal economies, they often fall short of fully articulating the cultural capital, economic capabilities of the informal processors, and economics of India's e-waste processing industry. For example, Turaga et al. (2019) calls for modifications in e-waste policy by recognizing the strength of the informal economy and the limits to scale economies in the e-waste industry. This high-level discussion, however, misses opportunities to integrate the cultural capital and economic capa-bilities into these policy recommendations. This fragmented view of India's informal e-waste processing context reinforces the status quo, and the prevailing suggestions for modifying the e-waste law continues to be based on overarching EU-based framework of EPR and the dominant discourse continues to be about *incentivizing* or *leveraging* or *upgrading* the informal economy to become part of the formal. Consequently, the policies developed without thinking through the informal economy's cultural capital, capabilities, and e-waste industry economics turns out to be ineffective, blocking their potential.

CONCLUSION

As we turn our attention to policy development, it is important to dispense with the claims that these homegrown informal e-waste management systems are an *alternative* way of doing business. The next chapter seeks to challenge the assumption that they should be *integrated* or *upgraded* or *leveraged* with the formal economy.[5] Any effective policy must acknowledge the strengths that existed before the turn to formalization in the 2000s, amplify those strengths, and work within the systems on the ground to design feasible and fruitful solutions. Until government and industry organizations fully recognize and unlock the potential of the unique network of human resources who play a central role in India's e-waste industry, policies will continue to waste resources and limit the effectiveness of economically, socially, and culturally viable enterprises.

The next chapter will dig into the policy context and suggest approaches to reforming policies not only in the context of India's e-waste sector, but in other contexts globally where colonial biases about *how business should be done* are preventing economically sustainable, culturally-appropriate innovations. Policies that intentionally unlock the potential of informal processors by leveraging their inherent economic capabilities have a greater chance not only of protecting their rich cultural capital, but also of reducing negative effects to health and environment, and advancing circular economy.

REFERENCES

Bourdieu, P. (1973). Cultural reproduction and social reproduction. In R. Brown (Ed.), *Knowledge, education, and cultural change: Papers in the sociology of education* (pp. 71–112). Taylor & Francis.

---

[5] The recent scholarly conversation on informal economy and their formalization in other contexts (outside the e-waste industry) echoes a similar perspective. Scholars have called for a shift in thinking on the informal economy and their formalization policies in the following ways:

- Formal systems to be integrated into the systems of informal economy and not the other way around (Chen et al., 2020).
- Formalization policies to be modified to unlock rather than blocking the potential in the informal economy (Guha-Khasnobis et al., 2006).

Britannica, T. Editors of Encyclopedia (2015, June 16). *Kolar gold fields*. Encyclopedia Britannica. https://www.britannica.com/place/Kolar-Gold-Fields

Chakrabarti, D. K., & Lahiri, N. (1996). *Copper and its alloys in ancient India*. Munshiram Manoharlal Publishers.

Chen, M., Carré, F., & Roever, S. (2020). Conclusion. In M. Chen & F. Carré (Eds.), *The informal economy revisited: Examining the past, envisioning the future* (pp. 257–268). Taylor & Francis.

Covey, S. R. (1989). *The 7 habits of highly effective people*. Free Press.

Doron, A., & Jeffrey, R. (2018, July 9). India's unofficial recycling bin: The city where electronics go to die. *The Guardian*. https://www.theguardian.com/cities/2018/jul/09/indias-unofficial-recycling-bin-the-city-where-electronics-go-to-die-moradabad

Granovetter, M. (1985). Economic action and social structure: The problem of embeddedness. *American Journal of Sociology, 91*(3), 481–510.

Guha-Khasnobis, B., Kanbur, R., & Ostrom, E. (2006). Beyond formality and informality. In G. Basudeb, R. Kanbur, & E. Ostrom (Eds.), *Linking the formal and informal economy: Concepts and policies* (pp. 1–18). Oxford University Press.

Harriss-White, B. (2003). *India working: Essays on society and economy*. Cambridge University Press.

Kenoyer, J. M., & Miller, H. M. L. (1999). Metal technologies of Indus Valley Tradition in Pakistan and Western India. In V. C. Pigott (Ed.), *The archaeometallurgy of the Asian old world* (pp. 107–152). University of Pennsylvania Press.

Lahiri, A. K., Banerjee, T., & Nijhawan, B. R. (1963). Some observations on corrosion resistance of ancient Delhi Iron Pillar and present time Adivasi iron made by primitive methods. *NML Technical Journal, 5*(1), 46–54.

Lam, A. (2000). Tacit knowledge, organizational learning and societal Institutions: An integrated framework. *Organization Studies, 21*(3), 487–513.

La Porta, R., & Shleifer, A. (2014). Informality and development. *Journal of Economic Perspectives, 28*(3), 109–126.

Murray, S. (2013, October 22). Waste: Precious metal shines light on recycling. *Financial Times*. https://www.ft.com/content/98235282-2b69-11e3-a1b7-00144feab7de

Polanyi, K. (1944). *The great transformation: The political and economic origins of our time*. Farrar & Rinehart.

Polanyi, M. (2009). *The tacit dimension* (2nd ed.). University of Chicago Press.

Reddy, R. N. (2011). *Specters of waste in India's "Silicon Valley": The underside of Bangalore's hi-tech economy* (Doctoral dissertation). University of Minnesota. University of Minnesota ProQuest Dissertations Publishing. https://www.proquest.com/docview/1024434302

Roy, T. (2009). *Traditional industry in the economy of colonial India*. Cambridge University Press.

Srikumar, S. (2014). *Kolar gold field: Unfolding the untold*. Partridge Publishing.

Srinivasan, S., & Ranganathan, S. (2013). *India's legendary wootz steel: An advanced material of the ancient world*. Universities Press.

Turaga, R. M. R., Bhaskar, K., Sinha, S., Hinchliffe, D., Hemkhaus, M., Arora, R., Chatterjee, S., Khetriwal, D. S., Radulovic, V., Singhal, P., & Sharma, H. (2019). E-waste management in India: Issues and strategies. *Vikalpa, 44*(3), 127–162.

Vaidyanathan, R. (2019). *Caste as social capital*. Westland.

CHAPTER 5

Policy Futures That Leverage the Strength of Informal Economy: Opportunities to Advance Circular Economy and UN SDGs

Abstract Policies that intentionally leverage the inherent capabilities of the informal processors and seek alignment with the way the e-waste industry is structured are more likely to succeed. This chapter proposes e-waste policy futures that leverage the strengths of informal processors by unlocking their potential and simultaneously mitigating negative health and environmental effects. The barrier to these policy futures has more to do with the systemic biases that have disenfranchised informal processors than with logistical challenges. By reframing the problem to understand the context of India's e-waste industry on a holistic level, the proposed policy futures accelerate the transition to achieving a fuller circular economy and sustainable development goals.

Keywords E-waste legislation · E-waste recycling · Informal economy · Extended producer responsibility · Circular economy · Informal recycling

T. S. Krishnan, *Untapped Knowledge in India's E-Waste Industry*, Palgrave Advances in the Economics of Innovation and Technology, https://doi.org/10.1007/978-3-031-50296-5_5

INTRODUCTION

In India's e-waste industry and in other industries, the informal economy[1] plays a vital role in provisioning essential goods and services. Informal processors have developed networks and systems that enable them to thrive, reaping economic rewards through approaches that honor and leverage cultural protocols. From the recycling of polyethylene terephthalate (PET) plastic bottles to bridging gaps in healthcare systems, informal economy plays a critical role in driving markets forward. In many cases, reframing the problem and minor policy adjustments offer opportunities to leverage the strengths within informal economy for better local and global outcomes. These benefits notwithstanding, many government policies, including the approach to e-waste management, reflect a mainstream reluctance to validate the business practices of informal economy. Often, these policies are rooted in long periods of colonization, extensive government intervention in business, and institutional inefficiencies (Austin et al., 2017).

As the previous chapters have shown, the systems that underpin and drive India's informal economy in the e-waste industry are made up of a complex set of economic and cultural factors like the capabilities that favor economies of scope, specialized social networks, and deep expertise. Policymakers around the world have implemented a Eurocentric approach to managing e-waste by utilizing the overarching framework of Extended Producer Responsibility (EPR). As we saw in Chapter 1, India's EPR-based e-waste policy is based on a model originating in Sweden, where there is virtually no presence of informal economy in waste processing. Research on the globalization of policies shows an increasing trend toward *fast policy*, the breakneck replication of policies around the world by reforming or transforming policies that work well in one country for application in others (Peck & Theodore, 2015). Indeed, fast policy has likely shaped policy significantly in the e-waste industry. As Peck and Theodore (2015) note, mimicking best practices isn't enough; there needs to be *translation* by taking into account the unique *local context*.

---

[1] In this chapter, the term *informal economy* includes the informal processors and the wider informal economy in e-waste (such as informal commodity traders and recyclers, and informal second-hand markets) and other industries. Sometimes, *informal economy* and *informal processors* are used synonymously in this chapter to contextualize the core ideas of policy futures.

This has not happened in the case of India's e-waste policy. The EPR-based e-waste policy is not tailored to the unique contexts of India with a *mainstream* informal economy in waste processing.

By building on local strengths, context-specific solutions have a greater chance not only of protecting rich cultural capital, but also of improving compliance and advancing agendas related to environmental sustainability. Such policies intentionally leverage the inherent economic capabilities of the informal processors and naturally align with the way the e-waste industry is structured. They support economies of scope, which are more appropriate for this context than economies of scale.

This chapter proposes e-waste policy futures that are meant to illustrate the positive outcomes that can emerge from the trajectories that take local human capital and innovative business strategies into account. These trajectories leverage the strengths of informal processors by unlocking their potential to simultaneously mitigate negative health and environmental effects, accelerate transition to a fuller circular economy, and support global objectives such as the UN Sustainable Development Goal 12.5—waste reduction through recycling and reuse. While the trajectories themselves are not meant to be scalable, the core features provide pathways for policy development in industry sectors where informal economy plays a vital role.

As we will see in this chapter, informal economy is not confined to e-waste industry alone. It plays a vital role in other industries such as plastic bottle recycling and primary healthcare, among many others. In these contexts, policy frameworks that harness the potential of informal economy to achieve sustainable outcomes are being proposed and practiced. Inspired by these policy interventions and the recent scholarly conversation, I propose policy futures explaining how reframing the problem can offer opportunities to leverage the strengths and tap into the potential of informal e-waste processors. The barrier to these policy futures has more to do with the systemic biases that have disenfranchised informal processors than with logistical challenges. By reframing the problem to understand the context of India's e-waste industry on a holistic level, the proposed policy futures have the potential to accelerate transition to achieving a fuller circular economy and sustainable development goals.

Unlocking the Potential of the Mainstream Informal Economy in Other Industries

The informal economy is not a *fringe alternative* or *residual*, but the *mainstream* in several major industries in India ranging from primary healthcare to PET bottles recycling. These examples demonstrate that informal economy exists outside e-waste processing and is a force to drive businesses and the economy forward. In the context of primary healthcare, Das et al. (2016) point out: *"Informal health care providers are the backbone of India's primary health care system."* More than 70% of India's qualified medical doctors work in urban areas serving only 30% of the population (Dagar & Dadhich, 2016). The acute shortage of trained medical professionals has led rural communities to depend on the support of informal healthcare practitioners. They do not have a formal medical qualification and provide 70% of primary care (Das et al., 2016). According to current policies, these healthcare workers are practicing illegally. If such laws were to be enforced, however, rural communities would have no access to healthcare at all.

Harnessing the inherent potential of informal healthcare workers would improve the healthcare system by addressing the acute shortage of formally trained doctors and leveraging the trust relationships that many informal healthcare workers have with their communities. When policymakers offer mechanisms through which informal practitioners have access to moderated forms of medical training, it can increase the supply of trained healthcare workers in rural communities. A recent academic study in West Bengal showed that training informal healthcare practitioners resulted in correctly diagnosing diseases and giving the right treatment at the right time as done by formally trained doctors (Das et al., 2016). Rather than considering informal healthcare practitioners as a problem to be eliminated, their inherent potential can be unlocked to improve healthcare delivery.

In addition to playing a critical role in India's primary healthcare, informal economy is critical to a range of other niches within the recycling industry such as PET bottle recycling. Over the last decade, India's PET bottle recycling industry has experienced the benefits of tapping into the potential of informal economy. A study conducted by scientists from the National Chemical Laboratory during 2016–2017 found that nearly 90% of the discarded PET bottles are reused or collected and recycled (NCL Innovations, 2017). Large beverage companies such as Coca-Cola and Bisleri and large recycling companies such as Arora Fibers partner with the informal economy to collect and process discarded PET bottles ("Coke, Bottler," 2011; Nayak, 2013; Patel, 2015; Vijayraghavan, 2010). Bisleri, which holds 60% market share of India's packaged drinking water industry, planned to sell the processed output, polyester fiber, to companies that make carpets, automotive parts, fabrics, road construction, etc. To make this venture successful, Bisleri partnered with members of the informal economy such as door-to-door waste collectors and scrap dealers.

Bisleri uncovered the inherent potential of informal economy through this partnership and improved the PET bottle recycling industry. Prior to the partnership with Bisleri, the informal economy never collected discarded PET bottles due to a lack of market. Bisleri offered attractive prices for the discarded PET bottles, thereby creating a market. Bisleri crafted this partnership by respecting the close-knit community and trust-based relations prevalent in the informal economy. This gave rise to a recycling market where the informal economy collected discarded PET bottles and sold to Bisleri for a market price. This worked out for Bisleri by opening up a new revenue stream and simultaneously protecting the environment by preventing landfill dumping of discarded PET bottles. This also helped the informal economy by increasing their income. Though there was a prevailing plastics law that mandated collection and recycling in the formal way, the recycling rates improved significantly only after the potential of informal economy was enabled to thrive. These large corporate companies are partnering with the informal economy to unlock their potential and move their overall business in a way that is aligned with their specific industries.

In both these contexts—primary healthcare and PET bottle recycling—we saw that the informal economy was the mainstream, and their potential was unlocked through appropriate practices that further helped to solve the problems related to healthcare access and plastic pollution, respectively. In a similar vein in the context of e-waste processing, informal processors are the mainstream, and their potential could be harnessed through appropriate policies that can help solve the negative environmental and health effects due to unsafe e-waste processing.

To develop these policies, we must frame the e-waste problem from a different angle. Currently, the problem is framed by only understanding the negative health and environmental effects. Thus, the prevailing policy solutions (EPR-based e-waste law) developed are intended to tackle only this aspect. As we saw in Chapters 1 and 2, the e-waste law has not achieved decision makers' intended outcomes and has resulted in business-as-usual practices. What if we could reframe the problem by understanding and appreciating the economic capabilities and cultural capital that enables informal e-waste processors to thrive and maximize *economies of scope*? This reframing of the problem can enable us to develop contextually-relevant solutions that preserve the inherent strengths of informal processors by simultaneously protecting their health and environment.

Rationale for Reframing the Problem to Unlock Informal Economy's Potential

Scholars have studied the informal economy and formalization policies in multiple industries such as loan provisions, forestry, street markets, and waste picking across countries such as Vietnam, Peru, South Africa, and India (Guha-Khasnobis et al., 2006). Collectively, these studies show that policymakers view the informal economy from two predominant perspectives (Chen, 2006):

1. A nuisance to be eliminated or regulated.
2. A vulnerable group to be formalized and integrated into the formal economy.

In the context of India's e-waste industry, the first perspective that frames the informal economy as a nuisance to be eliminated is not logically sound, nor is it inclusive. The existing working system that reuses and repurposes e-waste in useful ways is due to a thriving informal economy. It is hard to set up such a working system that naturally follows the core principles of a circular economy. The second perspective that frames them as a vulnerable group that needs to be formalized strips the informal economy of its strengths—economic capabilities and cultural capital—that are naturally aligned with the economic structure of the e-waste industry.

These two perspectives are dominant not only in policymaking circles, but also in the public opinion and popular media, too. The dominant media narrative being absorbed by policymakers often influences them to think about informal economy as a problem or nuisance to be eliminated. For instance, Gidwani and Corwin (2017) documented Corwin's visit to meet a government officer in India's environmental ministry department, where e-waste and other environmental laws are developed. On the officer's desk, there was a photocopy of an e-waste article published by a national newspaper. This article was being circulated throughout the department. It described the e-waste exports from developed countries to developing countries such as India and the toxic health impacts when this e-waste is processed by the informal economy. The government officer said that the prevailing EPR-based e-waste law was being tweaked to "*ensure loopholes are plugged*" and to "*prevent any 'leakages' to the informal sector*" (Gidwani & Corwin, 2017).

Recently, scholars have begun advocating for a third perspective that reframes the problem by leveraging the strengths of the informal economy and tapping into their potential (Chen et al., 2020). In the context of India's e-waste industry, this reframing of the problem translates to understanding what makes the informal processors thrive and how to leverage their strengths. The inherent economic capabilities—low cost, high flexibility, high quality, quick turnaround—of the informal processors are aligned with the economics of e-waste industry that is structured on economies of scope. Reframing the e-waste problem this way enables us to develop policy futures that leverage the strength of informal processors while protecting their health and environment.

POLICY FUTURES THAT UNLOCK THE POTENTIAL OF INFORMAL ECONOMY

Policy futures that make incremental changes offer opportunities to leverage the strengths of informal economy by aligning with the economics of the e-waste industry. Many of these potential interventions are low-barrier and low-cost options that would enable entrepreneurs in the informal economy to become co-creators of e-waste policy changes. The common thread of discussing the policy futures is to highlight the possibilities of potential scenarios that can unlock the potential of informal economy by facilitating them and strengthening their existing capabilities and cultural capital. What we need is informal economy-led policies, where the informal economy can play a central role facilitated by responsive public support, legislation, and resource allocation. From a policy perspective, this is an advantageous model because the health and environment are protected without sacrificing the strengths of informal processors. To develop specific targeted interventions, the government would need to consult with the informal processors. As Chen et al. (2020) noted: "The negotiations for such future policy frameworks should not only include informal economy but be led by them. A new cohort of policy makers are needed who are willing and brave to try out new approaches." By reframing the problem, the forward-thinking policymakers will be well equipped to devise targeted solutions led in partnership with the informal economy.

I propose three trajectories of policy futures that offer pathways for readers to develop new ideas or translate existing ones into targeted solutions. These trajectories are aligned with the existing strengths, business practices, and social networks of India's e-waste industry:

1. Facilitate markets to work.
2. Leverage opportunities to share resources.
3. Eliminate red tape that creates unnecessary barriers to growth.

FACILITATE MARKETS TO WORK

In the first trajectory, policies are centered around the needs of the mainstream informal economy, facilitating them to move their businesses forward while protecting the health and environment. The informal economy would be provided the agency to self-determine their business futures, join the decision-making table, and provide input along with other e-waste stakeholders. An example that has already been tested with the grain of this idea is the Quartz project. The Quartz project was initiated as a corporate social responsibility activity of IMU, an EU-based precious metal refiner, before the e-waste law was made binding in 2011. Stimulus, EU-based development agency, created a conducive framework by involving the government. They facilitated informal processors, a large formal processor EP, and IMU to do business with each other. Informal processors disassembled e-waste and took out circuit boards that were sold to the large formal processor EP, who then aggregated such circuit boards from multiple informal processors and exported to IMU for metal recovery. The benefit of this Quartz project was a reduction in negative health and environmental effects by diverting the hazardous metal recovery operations from informal processors to safe processing at IMU. Informal processors, EP, and IMU did business and protected the health and environment without any explicit e-waste law or policy. The grain of this tested idea could be scaled to develop policies centered around the needs of the informal economy.

Another potential scenario could be to install *micro-factories* in the informal processing hubs that would address prevailing concerns related to the environment and human health. This would enable informal processors to do metal recovery without using hazardous acids or being exposed to the toxic fumes. Micro-factories are low-cost, portable, small-scale technologies that extract several valuable metals from e-waste (Gough, 2016). They are called *micro* due to their small size when compared to the monolithic industrial-scale smelters used by metal refiners. For example, take the case of Veena Sahajwalla's laboratory research at the University of New South Wales. Sahajwalla's team has built portable small-scale smelters that can extract several metals from circuit boards, without the use of any hazardous chemicals and in a manner that is safe for human health and the environment (Gough, 2016).

Other potential scenarios emerged while presenting my research to an entrepreneur who is an expert in the international commodity markets and platform businesses.[2] This entrepreneur is one example of someone thinking creatively about potential scenarios by leveraging their expertise as they reframe the problem and learn more about this context. The entrepreneur suggested two approaches that leveraged the power of markets to unlock the potential of informal economy: (1) Set up dedicated India-specific commodities exchange for e-waste materials that would reduce the exposure of e-waste processors to the high price fluctuations in the international commodity markets. In turn, this could be leveraged to set minimum prices for those commodities that do not have a downstream recycling market. (2) Develop platform business models to reduce market frictions. The entrepreneur had noted the difficulty of newly-established formal processors to find appropriate buyers for the commodities disassembled from e-waste and was thinking of a web portal or a mobile app that could bring different players onto a common platform and facilitate exchanges between them. Similar to this entrepreneur, readers of this book could also reframe the problem and think creatively about potential approaches by leveraging their expertise.

---

[2] Platform businesses, such as Uber and Airbnb, bring producers (suppliers) and consumers (buyers) together, facilitate them to interact, and create value for each other (Van Alstyne et al., 2016).

LEVERAGE OPPORTUNITIES TO SHARE RESOURCES

In the second trajectory, policies leverage opportunities for resource sharing while protecting health and the environment. This enables the optimization of business models for economies of scope. By way of example, the idea of recycling parks offers a more effective approach to formalization by enabling government oversight without sacrificing an existing working system. A recycling park is a congregation of waste processors co-located in a common land. It operates using shared resources such as water and electricity to run facilities (Ehrenfeld & Gertler, 1997). In the context of India's e-waste economy, a recycling park could operate a power generation unit that provides electricity to all processors at a subsidized price. The processors can use a common effluent treatment plant that neutralizes toxic by-products of e-waste processing operated by the recycling park. In my fieldwork, both formal and informal processor stakeholders indicated that recycling parks with provisions for land and shared resources would be a viable solution to their current challenges. The Government of Karnataka proposed a 2,000-acre recycling park outside of Bangalore (Kumar, 2016). According to the current proposal, however, the park will be limited to formal processors, meaning that informal processors will not be able to access the services of recycling park.

The grain of this idea could be scaled to develop policies centered around the needs of the informal economy. By creating accessible spaces for informal processors to convene, trade materials, and process e-waste safely, the government would strengthen highly effective features such as economic capabilities, specialized social networks, and deep expertise that informal processors have developed. Modifications include scenarios where informal processors enter recycling parks with their e-waste, do the processing, do business with other companies at the park if needed, and exit with the output of their processing.

Take the case of the informal processor Aleph, who often uses the license of a formal processor to buy e-waste from global companies. Instead of disassembling the e-waste in informal ways, Aleph could visit the nearest recycling park facilitated by the government and disassemble the e-waste by using the dust collection equipment that absorbs the toxic dust emitted from e-waste during disassembly. Aleph would have complete ownership over the business and decide how to disassemble, whom to sell the materials to, and at what price. Such an approach would give informal processors a pathway to access dust collection equipment and protective gear such as face masks, mitigating the risk to workers of inhaling toxic dust and enabling processors to dispose of dangerous materials safely at the recycling park. In return for these services, the recycling park could charge a nominal fee to Aleph for this access. Recycling parks are one of many possible co-creative strategies to provide shared resources to facilitate lower operating costs while simultaneously mitigating the negative health and environmental effects.

ELIMINATE RED TAPE THAT CREATES UNNECESSARY BARRIERS TO GROWTH

In the third trajectory, policies eliminate the red tape that creates unnecessary barriers to the industry's growth—including formal and informal. One scenario is to provide Industry Status to e-waste processing. In India, Industry Status is a government classification recognizing e-waste processing as an approved industrial activity that receives special industry-specific incentives such as tax reductions, subsidies, financial assistance, and low electricity tariffs. This will help reduce the costs of setting up and operating a formal e-waste processing facility and encouraging further investments. Currently, it is hard for processors to receive these special incentives because *recycling* is not considered as an industrial activity in the country. An example that has already been tested with this idea is the Indian film industry. The film industry in India was given Industry Status in the year 2000. Filmmaking was included as an approved industrial activity. Until then, financing of films happened through private financiers who charged significantly high interest rates because banks refused to fund film production. Industry Status granted film producers and related businesses access to finance from legitimate financial institutions such as banks at low interest rates, venture capital, and private equity companies to fund film production, eventually enabling the industry to level up (Ganti, 2012).

However, as with other scenarios, this Industry Status scenario must be fine-tuned giving the center stage to the mainstream informal economy. In the context of India's informal processors, for example, any effective policy must account for cultural protocols related to Islamic finance. Islamic finance refers to financial transactions based on the principles such as interest-free lending that conform to Islamic law (The World Bank, 2015). Globally, Islamic finance assets are estimated at roughly

$3 trillion (Ahmed, 2022) and have proven to be an effective tool to finance economic development and reduce poverty in many Muslim and non-Muslim countries (The World Bank, 2015). The discussion on these adjustments and modifications must be designed in partnership with the informal economy. Only then, India can leverage the strengths of the informal economy and uncover its potential.

Across these trajectories and possible scenarios, the government must continue to play an important role in *facilitating* the e-waste processing industry. By truly understanding the unique economics of the industry and the potential of informal economy, decision makers at the local, national, and international levels can design better policies to support the public and the planet. These trajectories and scenarios are pathways to affordable, relatively minor adjustments with the potential to achieve large-scale impacts in India's e-waste industry.

BARRIERS TO POLICY CHANGES

The barrier to these policy adjustments has more to do with the systemic biases that have disenfranchised informal processors than with logistical challenges. By reframing our approach to understand the context of India's e-waste management system on a holistic level, we can leverage the untapped human resources and economic potential of the informal economy. The systematic bias against India's informal processors has emerged from a combination of Eurocentric pressures, formal processors advancing their agendas, and colonial mentality. Formal processors and Stimulus influenced the government to formulate e-waste policies to drive their own agendas. Stimulus is an Indo-German organization and they also had a Swiss collaboration during the phase when they were advocating for formalization of informal e-waste processors. Stimulus was influential in introducing India's EPR-based e-waste law. They produced written material and organized public events explaining the negative impacts of informal e-waste processing in India, from the early 2000s. They played a major role in drafting the EPR-based e-waste law by consulting the policymakers of government's environmental ministry, formal processors, and other stakeholders in the formal economy.

Formal processors, which would be less profitable than their informal counterparts on an even economic playing field, seized upon these opportunities to lobby for a competitive advantage. As we saw in Chapters 2, 3, and 4, informal processors can pay a higher price vis-à-vis formal processors for buying e-waste from bulk consumers due to their superior economic capabilities, specialized social networks, and deep expertise. This is not a level playing field for the formal processors. For them to survive in this industry and to compete with the informal processors, an e-waste law that would make it illegal to dispose e-waste to informal processors was essential.

Formal processors have a vested interest in diverting e-waste from informal processors, and they have leveraged policy, industry, and media platforms as well as capital to push this agenda. Formal processors influenced the government by writing in the mainstream media and speaking in industry events, to bring the EPR-based e-waste law. The formal processors entered this industry recently by investing a significant amount of capital and building large recycling facilities thinking in terms of scale economies. In the words of a government officer whom I interviewed during my fieldwork, these formal processors "*fought for*" bringing the e-waste law as "*they wanted to ensure that all bulk consumers give e-waste only to them.*" The officer specifically said: "*They wanted e-waste law to come. They wanted to eliminate the informal sector, so that they can be over them and do the business.*"

The portrayal of informal processors as a nuisance to be eliminated has been an effective strategy to diverting resources to formal processors. The global popularity of EPR-based e-waste law has advanced their interests, for example, by mandating producer brands of electronic products to collect and recycle discarded products or partner with formal processors for getting this done. Another important feature was that bulk consumers were mandated to dispose e-waste only to formal processors.

The government seems to take actions or make policies on the behalf of the formal processors. This phenomenon is related to the concept of regulatory capture (Stigler, 1971). Regulatory capture means the stakeholders who have high levels of business interest in the outcome of a policy are able to influence the members of policy making committee of the government to make policies that serve their interests (Stigler, 1971). In the context of e-waste, formal processors and NGOs, and development agencies such as Stimulus, influenced the government through writing, speaking, and publishing articles on informal e-waste processors and why the EPR-based e-waste law is the best possible way to manage negative health and environmental effects of informal e-waste processing.

Is it right to say formal processors had only *vested interests*? The answer is no. To some extent, it is true that the formal processors wanted a share of the business that was completely in the hands of the informal processors. But it is also true that at that time (during the early 2000s), the belief about the best way to manage e-waste was the EPR framework followed in the EU, and this belief continues to this day. It is quite natural that, to protect the health and environment due to the informal e-waste processing, the formal processors and the government began to *believe* that the EU-based framework is the most appropriate one rather than confidently studying one's own economy and developing context-specific solutions. Sadly, this colonial mentality is prevalent even now. The mentality that the Western ways are always better and superior is deep rooted (Ojha & Venkateswaran, 2022). There is a tendency to always look down upon the traditional systems and believe them to be inferior to the Western world (Ojha & Venkateswaran, 2022).

While on the one hand, it seems that policymakers in the government tend to copy environmental laws from EU, on the other hand it is also true EU makes significant efforts to influence the government to push their laws.

Scholars of international environmental law and politics have documented EU's interest in *exporting* its environmental laws around the world (Selin & VanDeveer, 2006, 2015a, 2015b; Vogel, 2012). This trend is visible not only in the context of e-waste, but also in industries such as automobiles and pesticides. This dissemination of Eurocentric policies affords the EU better businesses for European companies and becomes a maker of global laws that sets standards for the rest of the world to follow. It operates as a form of modern-day colonialism, influencing other countries to improve their domestic laws by copying the EU frameworks (Buck, 2007). For instance, in the context of India's e-waste industry, we saw the role played by Stimulus in advocating for formalization of informal e-waste processors. Additionally, e-waste quantity assessments in major Indian cities were jointly funded by EU. Funding came from the Indo-German partnership. These assessments quantified how much and what type of e-waste is generated in each city.

With this understanding, we can re-interpret the Quartz project, explained earlier, as one way to benefit IMU, an EU-based precious metal refining company. The Quartz project was initiated in 2010 as a corporate social responsibility activity of IMU. Stimulus worked with the government and facilitated informal processors, a large formal processor EP, and IMU to do business with each other. What were the events that happened before this project was initiated? In the early 2000s, Stimulus began e-waste awareness programs in India. By 2004, e-waste quantity assessment programs with EU funding were initiated in major Indian cities. This revealed the potential quantity and type of e-waste available from India. In the same year, IMU acquired the precious metal refining group of a major conglomerate for $800 million. In 2008, the top management of IMU revealed that they were not getting enough circuit boards to do precious metal recovery. In 2010, IMU, along with Stimulus, initiated the Quartz project to source circuit boards from India's informal processors. As we saw in Chapter 3, informal processors were doing the highest-value metal recovery by manually disassembling circuit boards and providing uncontaminated waste streams. This motivated IMU to initiate the Quartz project. Around the same time in the year 2010, India's e-waste law was announced that made informal processing illegal.

Based on this publicly available evidence and the scholarship on international environmental law and politics, we can see how EU affords better business for IMU by sourcing manually disassembled circuit boards from India's informal processors and influencing the government to introduce EPR that makes informal processing illegal.

By understanding and acknowledging these systemic biases, we are now well equipped to reframe the problem and leverage the untapped potential of informal processors.

Conclusion: Policy Futures Contribute to a Fuller Circular Economy and to Achieve SDGs

The policy futures, suggested trajectories, and scenarios recommended in this chapter are oriented toward realizing the potential of India's e-waste industry to develop as a circular economy. The Ellen MacArthur Foundation (2013) describes circular economy as *"an industrial system that is restorative or regenerative by intention and design."* We saw in Chapter 1 that such a circular economy existed in India's e-waste industry long before the EPR-based e-waste law was developed. If India's government and popular media understand how the *regulatory capture* has occurred in the e-waste processing industry and reframes the e-waste problem in a holistic manner, they will be well prepared to think about the contextual policy futures. By dismantling the existing dominant discourse and replacing that with one that leverages the strengths of the mainstream informal processors to unlock their potential, policy futures can accelerate India's transition to a fuller circular economy.

For instance, take the case of the recycling parks scenario suggested in the policy futures trajectory. These recycling parks that tap the potential of informal processors, through resource sharing, would increase the recycling levels in safe ways by accelerating the three core elements of a circular economy: *circling longer*, *cascaded use*, and *pure circles*. If such informal processor-friendly recycling parks were operational, more informal processors would process e-waste in these common spaces. The business ecosystem available in the parks could ensure that more materials such as glass could be sold to downstream commodity recyclers rather than wasting,[3] thereby accelerating *cascaded use*. The informal processors, buoyed by the legitimacy provided to them, would continue to use manual disassembly methods powered by their tacit knowledge to provide clean, uncontaminated waste streams, thereby accelerating *pure circles*. The shared resources provided in the parks could enable the informal processors and the formal processors to lower their operating costs, thereby enabling the processing of waste in a more profitable way. They would continue to reuse and repurpose e-waste to serve the needs of the downstream markets for second-hand products and commodity materials, thereby accelerating *circling longer*. Additionally, if these recycling parks are powered by renewable and low carbon energy sources, this means renewable energy is used to fuel the cycles of disassembly and reuse, thus achieving a fuller circular economy.

Transition to a fuller circular economy could further contribute to achieving sustainable development goals (Schroeder et al., 2019). The suggested trajectories and scenarios in the policy futures are intended to leverage the power of markets to increase reuse and recycling, thus reducing waste generation and simultaneously mitigating the negative health and environmental effects of unsafe e-waste processing. This translates to the waste reduction goal of SDG Target 12.5, which aims to

---

[3] In India's e-waste industry, nothing is wasted as long as the markets work. The glass materials disassembled by the informal processors in Bengaluru (southern part of India) can be sold to the glass recyclers in Delhi (northern part of India). But the transportation costs are so high that they erase the profit margins. This causes informal processors to waste the glass rather than selling them to a glass recycler.

"*substantially reduce waste generation through prevention, reduction, recycling and reuse*" (One Planet Network, n.d.). As we saw in Chapter 2, the current ways of measuring this target based on business-as-usual practices do not lead to actual results. The suggested scenarios in the policy futures are intended to alleviate the business-as-usual practices and thus help to truly achieve objectives such as SDG Target 12.5. For instance, the recycling parks ensure that all e-waste processed by the informal processors in the premises are done in ways that are safe to the health and environment. By accelerating *circling longer*, *cascaded use*, and *pure circles*, waste is reduced through increased levels of recycling and reuse.

The holistic approach described in this book can be used to re-examine our assumptions and look for culturally-appropriate policies in a wide range of recycling contexts. For instance, the sections on problem reframing and policy futures explained in this chapter can be usefully applied, for instance, to India's end-of-life vehicles recycling industry. In this industry, close to 100% of used vehicles are disassembled and reused by the informal economy (Chaitanya, 2016; TERI, 2022). Their operations are often unsafe for the health and the environment, and an EPR-based Eurocentric approach is being pushed as the only answer (Chaitanya, 2016; TERI, 2022).

Another recycling context is that of lead-acid batteries (LABs). My interviews with the stakeholders in the lead-acid batteries (LABs) recycling industry revealed that this industry also has the similar structure of e-waste processing context (Krishnan et al., 2016):

- A large presence of informal economy that reuse and recycle LABs in unsafe ways leading to negative environmental and health impacts.
- An EPR-based Eurocentric law.[4]
- Formal processors of LABs face stiff competition from informal processors.
- Pricing of LABs along the supply chain is dependent on virgin lead prices traded at the London Metal Exchange.
- Even after two decades of implementing the EPR-based batteries law, informal processors continue to persist, and many stakeholders such as bulk consumers and producer brands do not comply with the law.

---

[4] Batteries Management & Handling Rules is the EPR-based law that was introduced in 2001. The recent version was introduced in 2022 and is called Battery Waste Management Rules.

An industry veteran who owns a large-scale battery manufacturing and recycling facility said that LABs continue to be recycled in the informal economy, and it is difficult to solve this problem by forcibly closing down such facilities:

> Used batteries are not going to authorized recyclers. Unauthorized, informal recyclers are buying the used batteries. [...] Government should also keep one thing in mind. When they track and close down small informal recycling units, the problem is not getting solved. These people move to some other location and start new informal recycling units.

To resolve the problem of informal economies using unsafe methods in LABs recycling, we can use the holistic approach explained in this book and reframe the problem. This involves understanding economics, culture, and politics of LAB recycling and connecting them to explain the formalization experiences of informal LAB recyclers and the business-as-usual practices prevailing in the LAB recycling industry. This would help to develop contextually-relevant policies that leverage the strength and unlock the potential of informal economy by simultaneously minimizing the negative health and environmental effects.

While every emerging market includes its distinctive contextual factors, advantages, and challenges, there are key recurring themes that cut across both geography and specific markets. Many emerging markets are characterized by a mainstream informal economy whose practices are considered unsafe and unprofessional. Informal economies around the world face problem framing, wherein the public and policymakers perceive their *informal* status as an inherent problem and informal business owners as *vulnerable groups*. These challenges have led to legal and regulatory trends that force the informal economy to integrate with the formal. Ghana, for instance, has a robust emerging market for e-waste that is widely known as a global hub for e-waste processing. The country imports more than 200,000 metric tons of e-waste annually from developed countries (Lepawsky, 2018; Yeung, 2019). 90% of the imported and self-generated e-waste is processed by the informal economy, often in unsafe ways (Lepawsky, 2018; Yeung, 2019). The dominant narrative and policy thinking in Ghana is to consider the informal economy as a nuisance problem that needs to be eliminated or as a vulnerable group that need to be upgraded to the formal economy (Lepawsky, 2018).

By using the holistic approach followed in this book, we could embark on a study to understand the economics, culture, and politics of Ghana's e-waste processing context, and emerging markets more broadly. This will enable us to reframe the problem and develop relevant, grounded policies that leverage the strengths and tap into the potential of informal economy, simultaneously minimizing the negative effects. This would accelerate the transition to a fuller circular economy and achieve sustainable development goals.

REFERENCES

Ahmed, T. (2022, November 25). *What are the latest Islamic finance trends in the UK?* Refinitiv. https://www.refinitiv.com/perspectives/market-insights/what-are-the-latest-islamic-finance-trends-in-the-uk/

Austin, G., Dávila, C., & Jones, G. (2017). The alternative business history: Business in emerging markets. *Business History Review, 91*(3), 537–569.

Buck, T. (2007, July 9). How the European Union exports its laws. *Financial Times.* https://www.ft.com/content/942b1ae2-2e32-11dc-821c-0000779fd2ac

Chaitanya, S. V. K. (2016, November 26). To go green, law soon on recycling end of life vehicles. *New Indian Express.* https://www.newindianexpress.com/cities/chennai/2016/nov/26/to-go-green-law-soon-on-recycling-end-of-life-vehicles-1542771.html

Chen, M. (2006). Rethinking the informal economy: Linkages with the formal economy and the formal regulatory environment. In G. Basudeb, R. Kanbur, & E. Ostrom (Eds.), *Linking the formal and informal economy: Concepts and policies* (pp. 75–92). Oxford University Press.

Chen, M., Carré, F., & Roever, S. (2020). Conclusion. In M. Chen & F. Carré (Eds.), *The informal economy revisited: Examining the past, envisioning the future* (pp. 257–268). Taylor & Francis.

Coke, bottler tie up with rag pickers for recycling project. (2011, June 16). *The Economic Times.* https://economictimes.indiatimes.com/industry/cons-products/food/coke-bottler-tie-up-with-rag-pickers-for-recycling-project/articleshow/8864491.cms

Dagar, M., & Dadhich, S. (2016). *Healthcare in India: Current state and key imperatives.* KPMG. https://assets.kpmg.com/content/dam/kpmg/in/pdf/2016/09/AHPI-Healthcare-India.pdf

Das, J., Chowdhury, A., Hussam, R., & Banerjee, A. V. (2016). The impact of training informal health care providers in India: A randomized controlled trial. *Science, 354*(6308), aaf7384.

Ehrenfeld, J., & Gertler, N. (1997). Industrial ecology in practice: The evolution of interdependence at Kalundborg. *Journal of Industrial Ecology, 1*(1), 67–79.

Ellen MacArthur Foundation. (2013). *Towards the circular economy Vol. 1: An economic and business rationale for an accelerated transition.* https://ellenmacarthurfoundation.org/towards-the-circular-economy-vol-1-an-economic-and-business-rationale-for-an

Ganti, T. (2012). *Producing Bollywood: Inside the contemporary Hindi film industry.* Duke University Press.

Guha-Khasnobis, B., Kanbur, R., & Ostrom, E. (2006). Beyond formality and informality. In G. Basudeb, R. Kanbur, & E. Ostrom (Eds.), *Linking the formal and informal economy: Concepts and policies* (pp. 1–18). Oxford University Press.

Gidwani, V., & Corwin, J. (2017). Governance of waste. *Economic & Political Weekly, 52*(31), 44–54.

Gough, M. (2016, September 21). Saving silver: portable micro-factories could turn e-waste trash into treasure. *The Guardian.* https://www.theguardian.com/sustainable-business/2016/sep/21/saving-silver-portable-micro-factories-could-turn-e-waste-trash-into-treasure

Krishnan, T. S., Kokkattu, R., & Murty, L. S. (2016). *Take-back legislation for lead acid batteries: Stakeholders' perspectives in the reverse supply chain* (Paper Presentation). IIIE-POMS 4th International Conference on Best Practices in Supply Chain Management, Trivandrum, India.

Kumar, B. S. S. (2016, May 31). 2,000-acre waste recycling park proposed near Madhugiri. *The Hindu.* https://www.thehindu.com/news/national/karnataka/2000acre-waste-recycling-park-proposed-near-madhugiri/article8673308.ece

Lepawsky, J. (2018). *Reassembling rubbish: Worlding electronic waste.* The MIT Press.

Nayak, M. (2013, June 9). New age alchemists. *Business Today.* http://businesstoday.intoday.in/story/companies-that-are-making-wealth-fromwaste/1/195163.html

NCL Innovations. (2017). *PET recycling in India.* http://www.petrecycling.in/

Ojha, A. K., & Venkateswaran, R. T. (2022). Understanding the colonial roots of Indian management thought: An agenda to decolonise and theorise for Indian contexts. *Journal of Business Research, 149,* 700–712.

One Planet Network. (n.d.). *Target 12.5 waste reduction.* SDG 12 Hub. Retrieved January 26, 2023, from https://sdg12hub.org/sdg-12-hub/see-progress-on-sdg-12-by-target/125-reduce-waste-rrr

Patel, P. (2015, September 27). Fighting the PET recycling battle with fashion! *Daily News & Analysis.* https://www.dnaindia.com/lifestyle/report-fighting-the-pet-recycling-battle-with-fashion-2128987

Peck, J., & Theodore, N. (2015). *Fast policy: Experimental statecraft at the thresholds of neoliberalism*. University of Minnesota Press.

Schroeder, P., Anggraeni, K., & Weber, U. (2019). The relevance of circular economy practices to the sustainable development goals. *Journal of Industrial Ecology, 23*(1), 77–95.

Selin, H., & VanDeveer, S. D. (2006). Raising global standards: Hazardous substances and e-waste management in the European Union. *Environment: Science and Policy for Sustainable Development, 48*(10), 6–18.

Selin, H., & VanDeveer, S. D. (2015a). *European Union and environmental governance*. Routledge.

Selin, H., & VanDeveer, S. D. (2015b). Broader, deeper and greener: European Union environmental politics, policies, and outcomes. *Annual Review of Environment and Resources, 40*(1), 309–335.

Stigler, G. J. (1971). The theory of economic regulation. *The Bell Journal of Economics and Management Science, 2*(1), 3–21.

TERI. (2022). *Vehicle scrappage policy*. The Energy and Resources Institute. https://www.teriin.org/sites/default/files/2022-12/NTDC%20Discussion%20Paper%20on%20Vehicle%20Scrappage.pdf

The World Bank. (2015, March 31). *Islamic finance*. https://www.worldbank.org/en/topic/financialsector/brief/islamic-finance

Van Alstyne, M. W., Parker, G. G., & Choudary, S. P. (2016). Pipelines, platforms, and the new rules of strategy. *Harvard Business Review, 94*(4), 54–62.

Vijayraghavan, K. (2010, September 3). Bisleri boss Ramesh Chauhan plots new rags-to-riches story. *The Economic Times*. https://economictimes.indiatimes.com/bisleri-boss-ramesh-chauhan-plots-new-rags-to-riches-story/articleshow/6490425.cms?from=mdr

Vogel, D. (2012). *The politics of precaution: Regulating health, safety, and environmental risks in Europe and the United States*. Princeton University Press.

Yeung, P. (2019, May 30). The toxic effects of electronic waste in Accra, Ghana. *Bloomberg*. https://www.bloomberg.com/news/articles/2019-05-29/the-rich-world-s-electronic-waste-dumped-in-ghana

Index